1 MONTH OF
FREE
READING

at

www.ForgottenBooks.com

By purchasing this book you are eligible for one month membership to ForgottenBooks.com, giving you unlimited access to our entire collection of over 1,000,000 titles via our web site and mobile apps.

To claim your free month visit:

www.forgottenbooks.com/free1022254

ISBN 978-0-331-17061-0
PIBN 11022254

THE YEAR BOOK

OF THE

OIL, PAINT AND DRUG REPORTER,

A SYNOPSIS OF RESEARCHES AND DISCOVERIES IN

CHEMICALS, OILS, ETC.,

.DURING 1873.

CONTAINING BIOGRAPHICAL SKETCHES OF BARON
LIEBIG, C. H. LEONARD, FISHER HOWE
AND DAVID HOADLEY.

PUBLISHED BY

WILLIAM O. ALLISON,

42 CEDAR STREET.

NEW YORK:
THE REPORTER PRINTING ESTABLISHMENT,

CONTENTS.

CONTENTS.

ELECTROTYPED BY
CRUM & RINGLER,
NEW YORK.

W.G. Jackman

BARON JUSTUS VON LIEBIG.

AMONG the honored dead whose names give an historical interest to the necrology of the year 1873, none stood higher or filled a more honorable position than Baron JUSTUS VON LIEBIG. During a period of more than half a century he was an indefatigable laborer in the inviting field of organic chemistry, and his many and important discoveries and contributions to the current and standard literature of chemical and physiological science, entitle him to a rank only second to that of Humboldt. Probably the value of his labors will be more highly appreciated by future generations than by those which derive immediate benefit therefrom; but however this may be, later discoveries will only prove how important and fruitful of benefits to mankind, were the investigations and experiments by which LIEBIG laid the substantial foundations of the science of organic chemistry, especially as applied to agriculture. The greater part of his long and useful life was devoted to the study of the forces and elements which sustain life in plants and animals; and, with the resources of a well appointed laboratory at his command, he was enabled to solve many problems which had before been considered beyond the reach of science.

JUSTUS VON LIEBIG was born at Darmstadt on the 8th of May, 1803. His father had chosen for him the profession of pharmacy, and, after receiving a rudimentary education in the gymnasium of his native town, he was placed in the shop of an apothecary. The profession was not distasteful to him, but he was for some reason dissatisfied with the duties assigned him by his employer, and after a few months he returned to his father's house, and prepared for a course of study at the University of Bonn, which he entered in 1819. Completing the course at this institution, he continued his studies at Erlangen, and took a degree as Doctor of Medicine at the age of 19. In 1822 he received a travelling stipend from the Grand Duke of Hesse-Darmstadt, and with this assistance—more wisely bestowed than is usual with royal favors—he was enabled to visit Paris, where he remained two years. Here he studied with the famous Mitscherlich, devoting himself chiefly to the science of chemistry. While thus engaged his attention was especially attracted to those acids composed of carbon, hydrogen, nitrogen and oxygen combined with a base, known as fulminates. Whether the notice of the young chemist was directed to these dangerous substances by accident or by design, we do not know; but his experiments in this department of study doubtless shaped, in a great degree, his future life, as the elements he investigated, and with which he became so familiar in their natural combinations, are those with which organic chemistry chiefly deals. In after life he often alluded to his experiments with fulminates at Paris, as having furnished him a clue to the changes which the organic elements undergo in the compounds which form animal and vegetable tissues. These experiments were also more immediately beneficial to LIEBIG, for they brought him into notice as a young man of promise, and won him the favor and patronage of the greatest of contemporaneous investigators. In 1824 he was invited to read a paper on fulminates before the Institute of France, and accepted the invitation. Among the audience gathered to hear him was an attentive listener who, at the close of the paper, sought an interview and engaged in conversation. He made some inquiries concerning subjects not especially treated in the paper, asked LIEBIG what were his plans and prospects in life, and invited him to dinner. The stranger was a man of rather uncommon appearance, and LIEBIG, fearing to give offence, did not ask his name, supposing he could learn it from the janitor. That official did not know him, however, and LIEBIG, concluding that his new acquaintance was a person of small consequence, either forgot or neglected the engagement. A few days after he met a friend who asked why he had not dined with Humboldt, when a party of eminent chemists and men of science had been especially invited to meet him. It was too late to do more than apologise for his failure to respond to the invitation, but the acquaintance thus formed ripened into a strong friendship which was of great benefit to LIEBIG. Humboldt introduced him to Gay-Lussac and other eminent French chemists, and was instrumental in securing his appointment as Extraordinary Professor of Chemistry at Giessen in 1824. At the early age of twenty-one he entered upon these new and responsible duties, which he discharged with so much ability that, in 1826, he was made Ordinary Professor, and began the establishment of a laboratory for teaching practical chemistry. This was the first laboratory of its kind in Germany, and through it and its work the young chemist soon attracted the attention of scientific men throughout Europe. Here Hofmann, Will and Fresenius were educated for the great work they afterwards performed in promoting scientific progress, and here was founded the system of instruction, combining theory with practice, which has since been adopted in nearly all countries.

While thus engaged, LIEBIG began the series of contributions to the literature of science, which form so important a part of his life-work. In association with Wohler he began, in 1832, the publication of the *Annalen der Pharmacie*, a periodical still issued,. devoted chiefly to pharmaceutical chemistry. For many years LIEBIG was a frequent contributor to its columns, but of late years he was compelled to relinquish the work to other and younger hands. It was in 1832, also, that LIEBIG discovered hydrate of chloral, which, however, was not introduced to the medical profession until 1869, when Dr. Liebreich, of Berlin, first called attention to its value as a soporific. He was also the first to discover and describe chloroform, and one of the last acts of his life was to prepare a pamphlet establishing his claim to the discovery by incon-

testible proof. In 1838 he visited England for the first
time, and attended the meeting of the British Association
for the Advancement of Science, at Liverpool. Here he
read a paper on the composition and chemical relations
of lithic acid, in which he announced Wohler's great discovery of the composition of uria, and the method by which
it might be artificially produced. It is evident from the tone
of this paper that LIEBIG believed Wohler's discovery the first
step in a most interesting and hitherto unexplored path of
science, which would lead to the formation of other organic
substances and ultimately solve the problem of the chemistry
of life. LIEBIG's paper made a profound impression upon the
British scientists, and their appreciation of his talents is shown
by the fact that a vote was passed without a dissenting voice
inviting him to prepare and read two papers at the next meeting, one on isomeric bodies, and the other on organic chemistry.
The next meeting was held in Manchester, but LIEBIG was not
there, and it was not until 1840 that he responded to the invitation in another way, in the publication of his great work on
"Chemistry in its application to Agriculture and Physiology."

In the preface to this work the author states that he has
"endeavored to develop in a manner correspondent to the
present state of science, the fundamental principles of chemistry in general, and the laws of organic chemistry in particular,
in their application to agriculture and physiology; to the
causes of fermentation, decay and putrefaction; to vinous and
acteous fermentation and to nitrification. The conversion of
woody fiber into wood and mineral coal, the nature of poisons,
contagions and miasms, and the causes of their action on the
living organism, have been elucidated in their chemical relations." * From this brief resumé of the scope of the work, it
will be seen that LIEBIG ventured some distance into the realm
of speculation, as many of the processes and phenomena which

---

* Playfair's translation.

he claimed to have "elucidated in their chemical relations," are still unsolved mysteries. The work, however, was one of great interest and value, and did much to promote investigation and experiment in new directions. That portion of the work devoted to the action of poisons on the system is, perhaps, the most original and ingenious, and it shows throughout the power of a master-hand. He attributes the injurious action of poisons to two causes: First, by forming definite chemical compounds with the substances composing the flesh of the body poisoned, which render life impossible; second, by inducing chemical changes by contact, as is often seen in both inorganic and organic bodies. In the same way he explains the origin of the various forms of contagion, claiming that poisoning results from the introduction into the system of a substance capable of causing the solids and fluids of the body to assume a state of change similar to that in which the poisonous substance itself exists. This work excited a great deal of profitable discussion, the result of which was to show the author that where he had availed himself of the labors of others less thorough and accurate than himself, he had reached some mistaken conclusions; and in subsequent editions he made many and important changes. It was, however, instrumental in directing attention to the subjects it discussed, and in establishing the important truth that the careful and intelligent study of the physiology of plants is the indispensible prerequisite of progress in agriculture as an art. In 1855, LIEBIG published another work entitled "Principles of Agricultural Chemistry, with Special References to the late Researches made in England." Probably this is the most mature and complete work of this great student of nature's laws, and contains the best presentation which has yet been made of the relations between chemistry and agriculture.

In 1842, LIEBIG published his "Animal Chemistry, or Chemistry in its application to Physiology and Pathology," of which a third and greatly enlarged edition appeared four years later.

In this work he extended his researches to the laws of animal existence and development, and ventured the bold experiment of setting aside the hypothesis of a vital principle as accounting for the phenomenon of life—examining it as a physiological fact, the reason for which would be found by chemical analysis. By careful comparisons between that which is taken into the body and that which passes from it, he proved that the phenomenon of animal heat is the result of the combustion of carbon, and the group of proteinaceous compounds discovered by Muller were traced to their ultimate destiny in the production of animal tissue. It is impossible in the brief limits of a biographical sketch to enter upon the examination of the plan and scope of such a work. We can only say that it created a profound impression, and called out an amount of scientific research which has been productive of great and important benefits, and given us the science of physiology as we find it to-day—still with unsolved problems, it is true, but with a mass of truth based upon undoubted demonstration, which would never have been reached had a different method of study been followed from that begun by LIEBIG.

In 1849 another important work made its appearance, entitled "The Chemistry of Food," and the name of LIEBIG as its author secured it immediate and thoughtful attention. In this was given an account of his experiments on the changes which the tissues of the body undergo, and which result in the conversion of fibrine and albumen into gelatine and uria. In these experiments he demonstrated the universal presence in animal flesh of kreatinine, kreatine, lactic acid, phosphoric acid and inosinic acid; and of the existence of phosphate of soda in the blood, with the function of absorbing carbonic acid. He also called attention to the fact, not yet fully appreciated, that the proper cooking of food involves a correct understanding of the changes it undergoes in its preparation, and that with this knowledge a great improvement and economy is possible in the culinary art. Indeed, there is no department of human labor in which LIEBIG did not discover the chance for more rapid progress when chemistry shall have became a popular study. "For my own part," says he, in the preface to his "Chemistry and its Application to Physiology, Agriculture and Commerce," "I do not scruple to avow the conviction that, ere long, a knowledge of the principal truths of chemistry will be expected in every educated man, and that it will be as necessary to the statesman and political economist, and the practical agriculturist, as it is already indispensible to the physician and the manufacturer."

The great and varied services rendered by LIEBIG in the cause of scientific progress, naturally won for him honors of the kind which have been bestowed upon but few men with no other claim to royal favor than their knowledge gave them. His title of hereditary Baron was conferred upon him in 1845, by the Grand Duke of Hesse. His other honors consisted chiefly in the invitations extended to him to occupy the chair of chemistry in the principal universities of Europe, all of which he declined until, in 1852, he was induced to accept the professorship of chemistry in the University of Munich, with the position of President of the Laboratory. Here he remained, pursuing his studies and investigations, until his death, which occurred on the 18th of April, 1873.

In his personal and social relations LIEBIG was usually urbane and agreeable, but he could be severe and imperious when annoyed. He was especially intolerant of contradiction or criticism from ambitious men of small knowledge, who sought reputation by attacking him in the journals and in pamphlets, and when he consented to notice such criticism at all it was usually to come down upon the offender with crushing power. His severity in disputes made him to some extent unpopular, especially among the young men of the profession; but to any honest and industrious student of nature's laws, who sought knowledge rather than notoriety, he was always ready to extend a helping hand. He was, preeminently, a man of humane instincts, and his labors in the introduction of extract of beef, and experiments with methods of silvering glass had no other object than to procure a cheap and wholesome article of food for the poor and sick, in the one case, and in the other to rescue from disease and premature death the victims of mercury poisoning engaged in the manufacture of quicksilver mirrors. During the last few years of his life his mind was filled with benevolent schemes and projects, many of which he failed to realize. His knowledge was by no means limited to chemistry and kindred sciences. He was well read in the literature of the day, and could converse freely and intelligently on any topic likely to be a subject of conversation in the high circle in which he moved, and of which he was so bright an ornament. His principal recreation was found in playing whist—a game to which he was . very partial—and every summer he would make up a party of distinguished scientists and *savans*, to pass a few weeks in some quiet village, where he would endeavor to escape all attention and devoté himself to mental rest. Wohler, Clausius, Schoenbein, Andersen, and others of high rank in science and letters, were his companions on these pleasant excursions, from which all returned strengthened and refreshed to their congenial labors in promoting scientific progress.

Brief and unsatisfactory as a sketch of this kind must necessarily be, it recounts, we believe, the principal events of Baron LIEBIG's life, and it only remains to notice, in a general way, the practical results of his labors in the field of organic chemistry. The first and most important benefit of his discoveries was to direct attention to the importance of supplying to soil under cultivation the phosphates taken up by the plants. Against the system of farming which depends solely upon barnyard manure as a fertilizer, he waged relentless warfare, stigmatizing it as "vampire agriculture." He not only pointed out the evil of thus impoverishing the soil, but he discovered the remedy, and was the first to suggest the use of bones rich in phosphates as fertilizers. Experiment having shown that these resisted decomposition in the soil, he devised the method of treating them with sulphuric acid, and thus opened the way for the establishment of the business of manufacturing superphosphates. The results of this discovery was of immediate and permanent benefit. They doubled the turnip crop in England, and led to a rapid and sustained increase in the agricultural products of the Continent of Europe and the United States. Speaking of the employment of natural phosphates in Great Britain LIEBIG says: "What a curious and interesting subject for contemplation! In the remains of an extinct animal world, England is to find the means of increasing her wealth in agricultural products, as she has already found the great support of her manufacturing industry in fossil fuel—the preserved matter of primæval forests—the remains of a vegetable world. May this expectation be realized, and may her excellent population be thus redeemed from poverty and misery." He did not undervalue manures, but he considered them useful chiefly because of the mineral matter they contained, rather than because of their nitrogenous constituents. Notwithstanding this, he was a firm believer in the economy and advantage of utilizing sewage, and in more than one of his works on the chemistry of agriculture he draws a comparison between the thrift of China, supporting a dense population upon the products of her soil without importing any fertilizers, and of Holland and Alsace where the sewage is utilized; and the extravagance which necessitates the importations of vast quantities of fertilizing substances annually into England. In many other ways than those above noted his labors were productive of immediate benefit, but our space does not permit us to inquire in what directions, nor is the inquiry necessary to a just appreciation of the value of his life work. With him the end and aim of effort seems to have not only to do good, but to do it at once,—not only to be useful, but to be useful immediately. His works, especially those upon agriculture, were written for the people, and not for the man of learning, hence the simplicity of his language and the clearness of his argument. That he had reached many erroneous conclusions in the course of his varied and extensive researches, none who are at all familiar with his writings need to be told; but he was always ready to acknowledge an error and admit the truth, when subsequent study showed him the one or revealed the other. In his attempt to base the science of organic chemistry upon the hypothesis of component radicals, he was notoriously at fault, as was shown by Laurent and others; and in many other respects he was mistaken in his earlier conclusions; perhaps, also, in some of those reached at a mature age. In this respect, however LIEBIG was not alone among the students of nature. "Show me a man who makes no mistakes, and I will show you one who does nothing," is an epigram which originated with LIEBIG many years ago, but which has been accorded to nearly every one else but him. Few of those who have ventured far into unexplored realms of science have made so few and unimportant mistakes, and none have won a more permanent place in the hearts of the people of all the countries. Humboldt's name is known everywhere, even in countries where few can tell upon what his fame rests; Mayer, Bunsen, Kirchoff, Helmholtz, Faraday and many others are known, but LIEBIG is remembered as a friend, and his writings and precepts will be kept in mind as long as man shall draw sustenance from the soil.

SCIENTIFIC · CHEMISTRY.

Discoveries and Improvements during 1873.

THE ARTIFICIAL FORMATION OF ORGANIC SUBSTANCES.

By Dr. HENRY E. ARMSTRONG, F.C.S., Professor of Chemistry, London Institution.—From marsh-gas or methane, by a series of operations similar to those whereby ethane is converted into ethylic alcohol, an alcohol is obtained which proves to be identical with methylic alcohol, or wood-spirit, one of the main products of the destructive distillation of wood. These alcohols serve as the starting-points for the preparation of other hydrocarbons and alcohols, bearing relations to each other similar to those which obtain between methane and ethane, and between methylic and ethylic alcohols. Many of the alcohols, as prepared artificially, are identical with the alcohols which are obtained, together with ordinary alcohol, by fermenting saccharine substances, or which exist in the form of compound ethers in the ethereal oils extracted from various plants.

A long series of products is obtained by the oxidation of alcohol. This is first deprived of a portion of its hydrogen and converted into aldehyde, which latter is then converted by direct assumption of oxygen into an acid—ethylic alcohol, yielding acetic acid, the acid of vinegar. The other terms of the series of alcohols to which ordinary alcohol belongs are acted upon in like manner, and thus a series of aldehydes and acids is obtained. Many of these acids are identical with those which enter into the composition of the natural fats. The aldehydes are extremely alterable compounds, and readily undergo change, almost spontaneously in fact, being converted into bodies of more complex composition.

Formic aldehyde, the aldehyde of methylic alcohol, is probably formed in plants from the carbon and oxygen of the carbonic acid of the atmosphere (whence, as is well known, plants derive their carbon), and the hydrogen from water. One of the simplest transformations of formic aldehyde is its conversion into sugar; this conversion, however, has not yet been effected artificially, although formic aldehyde has been converted into a substance closely resembling the natural sugars. The aldehydes combine directly with ammonia, and the products readily part with the elements of water and are converted into alkaloids, one of which, that obtained from butyric aldehyde, is identical in nearly all respects with conine, the poisonous alkaloid of hemlock.

A variety of interesting derivatives is also obtained from the acids of the acetic series, such as glycocine, leucine. glycollic and lactic acids, all of which are substances found in various mineral fluids.

SULPHUR, SULPHURIC ACID, ETC., FROM GAS-LIME.

JULIUS KIRCHER, of New York, states that sulphur, sulphuric acid, and the sulphurets of sodium and potassium may be obtained from gas-lime by heating it to 300° F. in a closed retort, and passing steam at 600° F. over it, evolving sulphureted hydrogen, which passes to a leaden chamber, and is there supplied with air and ignited to produce sulphurous acid ; it is then mixed with nitric acid vapors, when the reaction produces sulphuric acid. The gas-lime is then mixed with clay, loam, or sand, and subjected to heat, when the silica or alumina unites with the lime and with oxygen, forming silicate of lime, etc., and liberating the sulphur. To produce the sulphuret of sodium or potassium, the gas-lime, etc., should be mixed with caustic soda or potassa, and allowed to stand until the reaction takes place.

ACTION OF SULPHUROUS ACID UPON INSOLUBLE SULPHIDES.

LANGLOIS having proved that alkaline sulphites are converted into hyposulphites by the action of sulphurous acid, another chemist named Guerout has repeated the experiment with the sulphides of other metals, and finds that the sulphides of copper, silver, gold, platinum, and mercury are not attacked. The sulphides of manganese, zinc, and iron readily dissolve in a strong solution of sulphurous acid, being at the same time converted into hyposulphites. The sulphides of cobalt, nickel, cadmium, bismuth, tin, arsenic, and antimony are slightly soluble, and undergo the same change into hyposulphites; varying quantities of sulphureted hydrogen are evolved, and sulphur separates. Further experiments, however, indicate that the sulphides are not converted directly into hyposulphites, but are first converted into sulphites which are afterward changed into hyposulphites.

This easy and rapid method of preparing hyposulphite of iron, zinc, etc., having been discovered, it remains to apply it to new and important uses, and such, we doubt not, will soon be found.

SULPHITE OF SODA AS AN ANTICHLOR.

THE term antichlor, which applied originally to any substance employed to destroy the free chlorine remaining in fabrics bleached with it, is now almost entirely limited to hyposulphite of soda, $Na_2 S_2 O_3$. During the reaction of this salt upon chlorine, free sulphur is deposited upon the fabrics, much to their detriment. The probable reason that this has never before been observed, is because its injurious effects have been attributed to over-bleaching. This finely divided sulphur, when deposited in the fibre of paper, gradually oxidizes to sulphurous and sulphuric acid, which renders the paper brittle, and, if written upon with iron ink, bleaches or fades it. This effect upon paper has sometimes been attributed to its containing too much wood-fibre.

A larger quantity of active sulphurous acid can be obtained from a given weight of sulphite of soda, $Na_2 SO_3$, than from an equal weight of the hyposulphite, and from this no sulphur is deposited, so that it ought most certainly to be preferred for use as antichlor on a large scale. We are informed by large manufacturers of chemicals that sulphite of soda can be made at a price not higher, in proportion to its efficiency, than the hyposulphite.

SOLIDIFICATION OF NITROUS OXIDE.

ACCORDING to Wills, nitrous oxide may be easily solidified by causing a rapid current of air to pass through the liquefied gas. Differing in this respect from carbonic acid, nitrous oxide may be kept liquid for some time in open vessels. Carbonic acid solidifies as soon as it escapes from its containing reservoir, because the tension of the vapor of the solidified acid, even at the moment of its formation, is considerably superior to atmospheric pressure; while liquid nitrous oxide attains—133° Fahr. and solidifies at—146°, so that the tension of its vapor is weaker than one atmosphere. The density of the liquid protoxide at 32° Fahr. is equal to 0.9004; its coefficient of dilatation is very considerable. It is insoluble in water.

MANUFACTURE OF ALKALINE CARBONATES.

HECTOR DE GROUSILLIERS, of Berlin, states that the alkaline carbonates may be manufactured from their haloid salts by introducing the latter, together with an alcoholic solution of carbonate of ammonia, into a closed vessel lined with lead and heated by steam, until a pressure of five atmospheres is obtained, whereby the alkaline carbonate is precipitated, the chloride of ammonia remaining in solution.

ACTION OF NITRIC ACID ON CHROMATE OF LEAD.

ON treating chromate of lead with about double its weight of nitric acid, a solution of chromic acid is obtained, accordidg to M. E. Duvillier, containing but two per cent. of oxide of lead. It is considered that the nitric acid decomposes the chromate of lead into chromic acid and nitrate of lead, which precipitates itself on boiling in presence of the excess of nitric acid employed.

A NEW MODE OF FILTRATION.

BY ISAAC B. COOKE.—A 300 c.c. flask is fitted with a perforated india-rubber stopper, through which a narrow glass tube of about six inches in length is passed for a distance of one inch. The lower end of the tube is drawn out to a fine point, and the other is slightly enlarged in form of a funnel. Into this funnel-shaped mouth a small quantity of cotton-wool is gently packed for about half an inch, and the remainder of the wool is left outside in the form of a spreading brush. To utilize this arrangement some water is poured into the flask, and the air expelled from the latter by steady ebullition; the stopper is inserted at the right moment, and, when condensation begins, the flask is inverted and the tube with its brush of cotton-wool plunged into the liquid to be filtered, the latter being contained in a small porcelain dish of about two and a half inches in diameter. When nearly all the liquid has been drawn up, a stream of distilled water may be directed upon the precipitate to wash it, and the washing can be repeated as often as necessary. The cotton plug is subsequently removed by forceps over the evaporating basin, and the end of the tube having been cleansed from adhering precipitate. the basin with its contents is dried over a lamp, ignited. and weighed.

HOW TO CLEAN GREASY VESSELS.

DR. WALZ has suggested a method for cleaning greasy beakers and photographic glass plates, which must at once commend itself to all practical chemists and photographic operators. He takes a dilute solution of permanganate of potash (kept on hand in a large stock bottle), to which a few drops of hydrochloric acid are added when used; and he pours in enough to wet the sides of the vessel to be cleaned. The greasy impurities are at once oxidized and removed. The method is preferable to the employment of bichromate of potash and sulphuric acid. The permanganate of potash solution can be saved and used repeatedly until, by the exhaustion of its oxidizing power, it ceases to act.

RAPID FILTRATION.

A SIMPLE contrivance, acting upon the same principle as Bunsen's filter, has been proposed by E. Fleischer. A wide-mouthed bottle is closed with a rubber cork twice perforated; into one of the perforations a funnel is fitted, while a short glass tube, bent at a right angle, is inserted into the other, and lengthened by means of a piece of rubber tubing with spring-clamp attached. The filter is capped with a small filter, then inserted and well moistened so as to rest against the funnel; afterward, the liquid to be filtered is poured upon it, and the air in the receiving-bottle rarified by sucking through the rubber tubing, which is then closed by the clamp.

CODÉINE.

THIS substance, which is also called *codea*, is beginning to find use in medicine, and some description of its properties as well as the method of its preparation may be of interest to some of our readers. Codeine, like morphine, narcotine, and several other alkaloids, occurs in opium. It was discovered by Robiquet in 1832. These alkaloids exist in opium in combination with certain vegetable acids, principally meconic acid. To obtain the codeine, it is first necessary to remove the meconic acid. An aqueous infusion of opium is evaporated to a syrup and mixed with a solution of chloride of calcium, which precipitates the acid as meconate of calcium, leaving the hydrochlorate of morphine and hydrochlorate of codeine in the solution, from which they crystallize if left to rest. These crystals are dissolved in water, and the solution, after purification with animal charcoal, is precipitated by ammonia, which separates the greater part of the morphine, leaving the codeine in solution. The filtered liquid is evaporated over a water-bath to expel the excess of ammonia, the morphine salt remaining in the solution being at the same time precipitated; the saline solution is concentrated and precipitated by caustic potash, and the precipitate of codeine is washed, dried, and dissolved in ether, whence it is deposited in crystals. From 100 lbs. of opium only six or eight ounces of codeine is obtained. Codeine is more soluble in water than morphine; it is also soluble in alcohol, ether, and ammonia, but it is quite insoluble in potash. It has a strongly alkaline reaction, restoring the blue color of reddened litmus, and precipitating the salts of lead, iron, copper, cobalt, nickel, etc.

The physiological effects of codeine resemble those of morphine in many respects. According to Robiquet, a dose of 0.3 or 0.4 of a grain produces in 24 hours, especially in an excitable person, a sensation of comfort and repose and a refreshing sleep, and a dose of from 1.8 to 2 grains produces heavy sleep with a feeling of intoxication after waking, sometimes also nausea and vomiting. More than three grains in 24 hours cannot be taken without danger of serious consequences.

CRYSTALLINE MERCUROUS IODIDE.

BY P. YVON.—By heating the required proportions of mercury and iodine in a closed flask on the sand-bath, at a temperature not exceeding 250 deg., the author has obtained mercurous iodide condensed on the upper part of the vessel in well-defined rhombic crystals, of a magnificent garnet-red color, which become yellow on cooling. If carefully heated, the compound sublimes unchanged; but if rapidly heated, the crystals melt at 290 deg. to a black liquid, which boils at 310 deg., mercury being given off, and a pale yellow crystalline sublimate obtained, which appears to be an oxyiodide.

ARSENIC IN SULPHURIC ACID.

IRON and copper pyrites (compounds of sulphur with the metal) are now extensively employed as sources of sulphur in the manufacture of sulphuric acid, and both these substances always or nearly always contain a notable quantity of arsenic. In order to separate the sulphur from the metal and convert it into sulphurous acid, the ore is roasted in suitable kilns and the gaseous and volatile products carried over, by means of a flue, into the leaden chamber where in the presence of other substances the formation of sulphuric acid is completed. But arsenic, like sulphur, is set free from the ore by the action of heat, and passes over along with the latter into the leaden chamber.

Arsenic is not only the most deleterious but the most difficult to remove of all the impurities of sulphuric acid. For medicinal purposes, therefore, and for use in certain medico-legal examinations, where perfect purity is required, an acid is employed made from pure sulphur, and even this has to be purified before it is fit for use. But in alkali manufacture, where immense quantities of the acid are consumed, that derived from pyrites is usually employed without previous purification, the products of the industry, such as hydrochloric acid, carbonate of soda, and sulphate of soda, being subsequently cleansed, when that is thought necessary. A recent writer on the subject, of much experience, Mr. H. A. Smith, strongly condemns this practice of first using an impure acid and then hunting up and removing the impurity after it has been distributed through a variety of products. "If the arsenic is to be removed at all" says Mr. Smith, "everything points to the sulphuric acid stage as that in which the removal ought to take place. Sulphuric acid is the cornerstone of alkali manufacture; cleanse it, and the whole is clean."

FLUORENE.

M. BERTHELOT announces, under the name of fluorene, a new and very fluorescent carburet contained in the portions of the tar of volatile oils between 300 deg. and 340 deg. C.

In order to extract the substance, instead of causing the portions of solid carburet which have passed the distillation between 300 deg. and 305 deg. C. to be crystallized in alcohol simply, a mixture of alcohol and benzine is used. By this means may be separated a small quantity of acenaphthene which remains in the mother liquor. The point of fusion of the mass, which is ordinarily 105 deg. C. after the first distillation and crystallization in pure alcohol, increases to 112 deg. after crystallization in alcohol mixed with benzine. The remainder of the purification consists in redistillation and crystallization in pure alcohol. The carburet possesses a quite pronounced violet fluorescence, which, however, disappears promptly on its being exposed to the light.

PURE SUB-IODIDE OF MERCURY.

LEFORT recommends the following method for preparing the sub-iodide of mercury free from iodine and from metallic mercury : 60 grains of pure crystallized pyrophosphate of soda are dissolved in 300 grains water, and 30 grains acetate of the suboxide of mercury added. The solution requires several hours. during which it is frequently shaken. If the soda salt is chemically pure. the mercury salt dissolves perfectly ; but this is seldom the case, and the excess of alkali precipitates some oxide of mercury, so that the solution requires filtering. It is then still further diluted with water, and a solution of 30 grains iodide of potassium in two ounces of water gradually added with constant stirring or shaking. This produces a precipitate which is at first a brownish green, but becomes a bright green, closely resembling oxide of chromium, and on settling acquires a yellow-green color. If the mercury solution contains any mercuric salt at the start, some biniodide of mercury is formed, giving the liquid a pinkish color; but this is easily avoided by adding a slight excess of iodide of potassium, which is so dilute as not to decompose the sub-iodide, while it is able to dissolve the biniodide. The precipitate is washed with cold water by decantation, collected on a filter, and dried with gentle heat in the dark.

OZO-BENZINE—A NEW EXPLOSIVE.

MM. HOUZEAU and Renard state that, by causing concentrated ozone to react upon pure benzine boiling at 178 deg. Fah., a solid body is formed of gelatinous appearance, to which the name of ozo-benzine has been given, The formula is $C^{12}H^6$ ($C_2 3 H 6$). Dried in *vacuo*, the substance becomes solid, white, amorphous, and highly explosible. It detonates with violence under the influence of shocks or heat. A few grains exploded in the laboratory shattered the glass in the windows. It is very unstable, and when left either in air, carbonic acid, or even in a vacuum, it changes rapidly. Among the products of the aqueous decomposition of ozo-benzine, is noted the presence of acetic and formic acids, and also that of a solid acid, very soluble, becoming colored brown by potash or soda. Another composition is also formed which has an agreeable odor and no acid reaction.

PURIFYING CARBONIC ACID GAS.

THE impurities contained in carbonic acid gas when derived from the combustion of coke, coal, etc., may be removed, according to Asa P. Maylert, of New Britain, Ct., by passing the gas, as it issues from the furnace, through or over a mixture consisting of deutoxide of lead held in suspension in water, or a weak solution of acetate of lead, or of some other soluble salt of lead.

ANILINE FOR PRINTING BLACK.

THE degree of purity of commercial aniline, says the *American Chemist*, is of the greatest importance in the manufacture of different colors, and especially of blue and black. As aniline black is developed by printers themselves and not bought ready for use, the following test will enable them to determine the quality of the article they have to use:

Any aniline oil which does not boil under 192 deg. C. must at once be rejected; and the nearer its boiling-point is to that of pure aniline, 180 deg., the finer will be the black color produced. For practical tests, several methods may be followed. Beaume's areometer gives some indication of quality. Any aniline of from 20 deg. to 30 deg. B. always gives a black color if not fraudulently adulterated. If heavier, it generally contains undecomposed nitro-benzol; if lighter, too much toluidine. Fractional distillation gives a more reliable result. The percentage of aniline distilling between 180 deg. and 185 deg. C. represents the true value of the article. Concentrated sulphuric acid diluted with three times its weight of water is also a good test. About one part of aniline is mixed with at least three parts or the dilute acid; a thick paste of sulphate of aniline is formed, and more water is added to dissolve the salt, when any tarry impurities and also nitro-benzol collect at the top.

The quantity of aniline oil used is enormous, being, in 1869, 3,500,000 pounds, of about 10.000 pounds per day. Of this, Germany took two million pounds, and the rest was divided between Switzerland, England, and France. The quantity of coal which must be converted into gas to furnish sufficient benzol for 3,500,000 pounds of aniline is astonishing. It is estimated that 1,600 tons of coal will produce one ton of aniline. Three and a half million pounds or 1,600 tons of aniline require therefore 2,500,000 tons of coal, which, in the first instance, would give 25,000,000,000 cubic feet of gas.

BENZINE-POTASSIUM, A NEW EXPLOSIVE.

BY H. ABELJANZ.—When benzine and potassium are heated together in a sealed tube to 240 deg.--250 deg., they combine and form *benzine-potassium*. It is a black mass which appears blue when viewed in thin layers. In the dry state it is very explosive, and it is decomposed violently by water. The gradual action of water on it gives rise to diphenyl.

FORMATION OF CRYSTALLINE ANTIMONY.

THE antimony is deposited on copper from a hydrochloric acid solution of antimonious chloride. On removing it, and pulverizing it in a mortar, it detonates, and is changed from the amorphous to the crystalline condition. The author has not observed if this result is produced by heat, as is the case with rhodium and iridium.—W. R. *Dingl. Polyt. Jour.*, ccvii. 427.

PRODUCTION OF FURFUROL.

By C. Greeville Williams.—When fir-wood is heated with water in a closed vessel, and a pressure of 100 lbs. to the inch maintained for a considerable time, an acid·liquor is obtained, from which, by repeated rectification, two principal products may be separated. These are methylic alcohol and an oil, (furfurol,) the yield of the latter from 100 lbs. of wood being 10 ounces.

A mean of four experiments showed that 100 parts of crude oil gave 39.3 of furfuramide, corresponding to 42.3 per cent. of furfurol. Hugo Muller corroborates the foregoing results. Having operated in a similar series of experiments with bamboo-wood, he obtained an appreciable quantity of a heavy oily liquid, which exhibited all the properties of furfurol, and produced, on the addition of ammonia, the well-characterized furfuramide. Muller does not think that the production of furfurol is due to the action of the organic acids which are simultaneously or previously formed ; at the same time he mentions that when the wood is treated with caustic soda-lye instead of water, the formation of furfurol does not appear to take place.

Furfurol may also be obtained by the action of high-pressure steam upon wood. The apparatus used consists of a vessel capable of withstanding a pressure of 500 lbs. to the square inch, within which is placed a cylinder of perforated metal filled with sawdust and standing on a perforated shelf ; the bottom of the vessel is covered with a layer of water not reaching to the shelf. The vessel having been tightly closed, is heated in an oil-bath to 198 deg. C. for three or four hours, after which it is left to cool till the pressure has completely gone down. It is then opened and connected with a condenser, and heated till about three-fourths of the water present has distilled over. The watery distillate thus obtained smells strongly of furfurol, and when mixed with ammonia, yields crystals of furfuramide.

It appears, then, that furfurol may be produced from wood by the action either of water or of steam at high temperatures and pressures. The author finds, however, that it is not produced by distilling sawdust with water under the ordinary pressure.

PREPARING AMMONIA SALTS.

Bobrownicki. of Paris proposes to prepare ammonia salts from the ammonia liquor of gas-works, by acidifying and then treating it with fluoride of silicon, chloride of silicon, hydrofluor-silicic acid, or an alkaline silicate. The silicon compounds carry down the suspended bodies, and those in solution, and hold them in a solid or half-solid form. Bobrownicki calls the precipitate a silicoid. It furnishes the crude material for preparing ammonia salts in the usual manner.

A NEW ACID — TAURO-CARBAMIC ACID.

By H. SALKOWISKI.—The author has already shown that when taurin is administered to the human subject, a portion passes unaltered into the urine. He now finds that the greater part of the taurin becomes transformed into an acid containing the elements of one molecule of taurin and one molecule of carbamic acid minus one molecule of water.

The new acid, *tauro-carbamic acid*, $C_3H_8N_2SO_4$. is obtained by precipitating the urine with acetate of lead, removing the lead from the liquor by hydrosulphuric acid, evaporating the filtrate, and adding alcohol to the concentrated solution. The crude sodium salt which then separates must be treated with animal charcoal and afterward decomposed by sulphuric acid. The crude acid may now be extracted by alcohol, which leaves it, on evaporation, in the form of a syrup which must be freed from sulphuric acid by means of barium hydrate, and from chlorine by means of silver carbonate.

The researches of Schuitzen render it probable that tauro-sulphamic acid is also excreted in the urine when taurin is exhibited, and the author finds that normal urine contains a trace of tauro-carbamic acid.

It has also been noted that a mixture of bicarbureted hydrogen and ozone detonates violently without the action of light, heat, or electricity. The ozone must be strongly concentrated.

NEW COMPOUND OF BICHLORIDE OF MERCURY AND BROMIDE OF POTASSIUM.

IF to a solution of bromide of potassium be added an equivalent proportion of bichloride of mercury, in powder, the latter salt dissolves quite readily, and a solution is formed which, when evaporated, affords needle-shaped crystals, consisting, apparently, of a definite compound of the two salts; or it may be that an interchange of elements takes place, and that the new salt is composed of bromide of mercury and chloride of potassium. The solution deposits crystals of the same form even when evaporated to the last drop. These crystals are permanent in the air, and when just removed from the mother liquor are transparent, but, on drying, become white, with a nacreous appearance. The salt dissolves readily in water. It gives with iodide of potassium, a scarlet precipitate of biniodide of mercury: with nitrate of silver, a precipitate of bromide: with chlorine water, followed by chloroform, a solution of bromine in the latter liquid.

DETERMINING ANILINE COLORS.

THE *Chronique de l'Industrie* describes a new method, the principle of which consists in the fixing of the coloring matter to be tested on a plate of glass, by means of collodion. The thin film thus obtained is compared with another of the typical coloring material prepared in the same manner.

ANTHRACENAMINE.

BY T. L. PHIPSON.—This new base may be obtained by adding anthracene in powder, in small quantities at a time, to ordinary nitric acid contained in a capsule which can be cooled if necessary. A soft reddish-brown mass is obtained which melts easily, and can be drawn out into long golden-yellow filaments. This product contains a certain quantity of mononitranthracene, $C_{14}H_9NO_2$, soluble in alcohol, from which it crystallizes in small yellow needles. If the temperature is allowed to rise and the acid boils, several other products are obtained and much oxanthracene. The product is washed and placed in a flask with tin and hydrochloric acid, diluted with its own volume of water and boiled quietly for an hour, then filtered. The filtered liquid contains chloride of anthracenamine and tin chloride; the base is extracted by excess of potash, which dissolves the tin oxide and leaves the anthracenamine. It is necessary to repeat the operation twice to get rid of all the tin.

Anthracenamine is a pale yellow powder, forming soluble and crystallizable salts with sulphuric and hydrochloric acids. It is very soluble in alcohol, but slightly soluble in water; its odor is very slight, and its taste is hot, pungent, and persistent, very like that of the unknown substance which exists in the arum maculatum. Its acid salts produce with potassium bichromate a characteristic emerald-green color, and finally precipitate a powder of this color which is soluble in alcohol. This solution presents no marked peculiarity when viewed in the spectroscope. The reaction is as characteristic of anthracenamine as the blue color produced in similar circumstances is of naphthylamine. It is not obtained, however, with lead peroxide or with calcium hypochlorite, but it is obtained with concentrated nitric acid.

Anthracenamine is easily decomposed; and from the percentage of nitrogen it contains, the author concludes that its composition may be represented by the formula, $C_{14}H_{11}N$.

A NEW RED FROM ANILINE.

By F. HAMMEL.—If sulphur chloride be added to 20-25 grammes of aniline, with constant agitation, a red solid product is obtained almost immediately. On treatment with acetic acid and filtration, a red solution is produced, from which the coloring matter is deposited on evaporation as a black mass, soluble in acetic acid, alcohol, and ether. On addition of water to either of these solutions, a gray precipitate is thrown down.

Artificial cryolite may be made by taking crude, or better, distilled hydrofluoric acid containing five per cent of anhydrous acid and half saturated with pure alumina. A solution of sodium chloride is then added until the mixture contains three equivalents of soda for one of alumina. The precipitate is artificial cryolite.

FREEZING-POINTS OF ACETIC ACID AND WATER.

BY E. GRIMAUX.—The addition of water to glacial acetic acid lowers the freezing-point down to a certain limit, beyond which the further addition of water raises it. The author has endeavored to ascertain this limit by experiment. He introduced mixtures of acetic acid and water (in proportions determined by weighing) into a small test-tube furnished with an alcohol thermometer, the zero-point of which was verified every day, and placed the arrangement in a freezing mixture. When the thermometer had sunk some degrees below the expected freezing-point of the mixture, the test-tube was shaken to cause solidification of its contents, and the highest point to which the thermometer rose at the moment of crystallization was read off. The standard acetic acid employed solidified at 14.4 deg., and hence, according to Rudorff's tables, contained 1.25 per cent. of water.

The lowest freezing-point of the solution obtained was 24.1 deg. which corresponds to a mixture containing 37 or 38 per cent. of water, that is, to an acid represented by the formula $C_2H_4O_2$ plus $2H_2O$.

PHOSPHORIC ACID.

THE occurrence of phosphorus in combination with the ores of iron has long been an annoyance to iron manufacturers, and many rich ores are worthless from the phosphorus, which makes the iron brittle and useless. Julius Jacobi proposes a method of freeing iron ores from phosphorus, and at the same time saving the phosphoric products for agricultural purposes. His process consists in roasting the ore and crushing it, and, after placing it in a proper receiver, submitting it to the action of water charged with sulphurous acid under pressure. The ore is then washed with water, to remove all the soluble products, and the phosphoric acid. precipitated from the water with fresh-burnt lime, is obtained as a neutral phosphate of lime. If effectual and not too expensive, the proposed method is very important, as rendering many ores available which are now regarded as worthless, and at the same time supplying a demand in agriculture which has heretofore been but imperfectly met.

DETERMINATION OF MANGANESE IN CAST-IRONS AND STEELS.

ONE of the main causes of the loss of manganese in these determinations springs from the employment of too large an amount of acetate of soda in the previous precipitation of the iron. The author finds that in a perfectly neutral solution 1 grm, acetate of soda suffices to precipitate completely 1.1 grm of iron in 500 C.C. of solution, and even in the presence of 1 grm. of acetic acid. When the determination of the manganese alone is required, he cools the liquid, makes up its volume to 500 C.C., filters through a dry filter, and determines the manganese in 250 C.C.—*Kiesser.*

TEST FOR ARSENICAL COLORS ON WALL PAPERS AND IN PAPER GENERALLY.

PROFESSOR HAGER recommends the following method for detecting this dangerous class of arsenical colors, which, we may remark, are not confined to green alone, for even red sometimes contains arsenic : A piece of the paper is soaked in a concentrated solution of sodium nitrate (Chili saltpetre) in equal parts of alcohol and water, and allowed to dry. The dried paper is burned in a shallow porcelain dish. Usually it only smoulders, producing no flame. Water is poured over the ashes, and caustic potash added to a strongly alkaline reaction, then boiled and filtered. The filtrate is acidified with dilute sulphuric acid, and permanganate of potash is added slowly as long as the red color disappears or changes to a yellow brown upon warming, and finally a slight excess of chameleon solution is present. If the liquid becomes turbid, it is to be filtered. After cooling, more dilute sulphuric acid is added and also a piece of pure clean zinc, and the flask closed with a cork split in two places. In one split of the cork a piece of paper moistened in silver nitrate is fastened, in the other a strip of parchment paper dipped in sugar of lead. If arsenic is present, the silver soon blackens. The lead paper is merely a check on the presence of sulph-hydric acid. According to Hager, the use of permanganate of potash is essential, otherwise the silver paper may be blackened when no arsenic is present.

COMPOSITION FOR GAS.

W. H. STERLING, of San Francisco, proposes the manufacture of gas by distilling in a retort blocks made as follows : To crude petroleum or coal-tar add, in proportion to its richness or gravity, ten to twenty per cent. of water, and from one-half to two per cent. of caustic lime. Thoroughly agitate the whole to produce an emulsion, and then mix sufficient finely-pulverized ashes to make a plastic mass, which may be moulded into block suitable for placing in a retort.

ACTION OF ETHER UPON IODIDES.

BY E. FERRIERE.—When a concentrated solution of an iodide is mixed with starch-paste and then shaken with ether, part of the iodine is separated, and the starch is turned blue. If the solution is dilute, the bluing does not appear till after two or three hours, and in extremely dilute solutions, not till after two or three days. On filtering from the blue starch and adding more ether, a blue color is again produced, and so on, till at length all the iodine is removed from the compound. Mineral waters containing iodides exhibit, when thus treated, the same reactions as artificially prepared solutions.

DYEING OF FELT WITH ANILINE COLORS.

FOR the dyeing of felt hats, aniline colors can be used in every case. The coloring matter is used repeatedly to make the tint satisfactory. If the dyeing follows the fulling, the felt is not penetrated so easily, but the hair can be directly dyed and the dyed hair fulled. For this purpose, a solution of the dye is made in boiling water, then allowed to cool, and filtered. A pan with water heated to 30 deg. is prepared, and into this the necessary quantity of dye is introduced, stirred up, and the hair moistened, and inclosed in a basket is placed in the bath. The bath is repeatedly heated to 60 deg., and the basket agitated therein continually. Fresh coloring matter is introduced when the hair has absorbed a certain amount, the basket being for the instant removed.

When the hair is fully dyed, the basket is removed and the hair allowed to cool, and it is then well rinsed. Mixtures of aniline colors may be used for particular tints with good effect.

For *brown*, the by-products from fuchsine are employed, which are known in the trade as "cerise," "merron," etc. These give with indigo-carmine and picric acid, with addition of a little sulphuric acid, splendid brown shades. For the preparation of the favorite "B'smarck," a solution of Manchester brown can be used, which is toned down by addition of indigo-carmine, picric acid, and fuchsine.

RUBIDIUM FROM BEET-ROOT ASH.

BY C. PFEIFFER.—The mother-liquor left in preparing saltpetre from the ash of beet-root molasses is mixed with saw-dust and deflagrated, the charred mass exhausted with water, the solution evaporated, and the sulphates and chlorides allowed to crystallize out. This second mother-liquor is mixed with hydrochloric acid, heated, filtered from precipitated sulphur, etc., and boiled with nitric acid till all iodine and bromine are expelled. The rubidium is now precipitated from the diluted solution by platinum tetrachloride, and separated as usual.

Direct experiment shows that one kilogramme of the beet-ash of Northern France contains 1.75 grm. of rubidium chloride, and that the rubidium chloride is to the sodium chloride and potassium chloride as 1 : 126 and 331.

PURIFYING QUICKSILVER.

MERCURY, when being manufactured, frequently becomes foul from the mixing with it of unreduced cinnabar the soot of the material employed in reducing it, and other impurities; but it is stated by Messrs. Randol & Wright, of New Almaden, Cal., that, by treating it under agitation with heated water and alkaline matter, the foreign matter will unite with the water, etc., allowing the mercury to be drawn off clean and bright.

AURINE.

By R. S. DALE, B. A., AND C. SCHORLEMMER, F. R. S.—Kolbe and Schmitt obtained in 1861 a red coloring matter by heating phenol with oxalic acid and concentrated sulphuric acid. Since that time, this color has been largely manufactured, and is found in commerce under the name of *aurine, yellow coralline,* or *rosolic acid.* The latter name, as is well known, was first given by Runge to a red body which he obtained from coal-tar, and this name was afterward employed to designate all red compounds which may be obtained from phenol by different reactions.

We refrain from giving a historical sketch of these red phenol-colors, because there cannot be any doubt that, according to the mode of preparation, different compounds are formed.

The analysis of the crystallized body, which Fresenius calls *coralline,* gave numbers agreeing with the formula $C_{40}H_{38}O_{11}$. We shall have to refer to Fresenius's paper again, and will only for the present mention that his coralline is not identical with the compound which we have obtained, and for which we retain the name of *aurine,*

Commercial aurine is a brittle, resinous body, having a beetle-green lustre, and yielding a red powder.

The purification is easily effected by adding concentrated aqueous, or, better, alcoholic ammonia, to a cold, concentrated, alcoholic solution of crude aurine. A crystalline precipitate, a compound of aurine with ammonia, separates out, while the other bodies contained in the crude product remain in solution. The ammonia-compound was washed with alcohol by means of the filter-pump; after drying, it forms a dark red crystalline powder with a bluish lustre. It is a very unstable body, losing its ammonia completely when it is exposed to the air for some time.

By boiling it with dilute acetic acid or hydrochloric acid, aurine is obtained as a crystalline, brownish-red powder having a green lustre; it must be purified by repeated crystallization from acetic acid.

By the first crystallization, it was generally obtained in small, dark red needles, with a steel-blue reflection; afterward it crystallized in larger needles or prisms, having the color of chromic acid and a brilliant diamond lustre, or of a darker shade showing a blue or greenish-blue reflection, and once we obtained it in small crystals having the beetle-green lustre of the salts of rosaniline.

The finest crystals were formed by the spontaneous evaporation of an alcoholic solution containing acetic acid. We have analyzed these different specimens partly dried at 100 degs. and partly at a higher temperature, and although samples of the same preparation gave very concordant results, those of different preparations varied very much in their composition. We found that the reason for this was that aurine most obstinately retains water and acetic acid, which, however, as we believe, are not chemically combined with it.

From hot concentrated hydrochloric acid, aurine crystallizes in slender red needles, which. when dried at 110 degs., still retain a large quantity of hydrocloric acid. We tried to obtain the pure compound by precipitating a dilute alkaline solution of aurine with weak hydrochloric acid and washing the precipitate with the filter-pump. but the product thus obtained also contained hydrochloric acid when dried at 110 deg.

By the spontaneous evaporation of an alcoholic solution, aurine is obtained in dull red crystals with a green lustre, which when dried at 110 degs. do not contain any alcohol. but several per cent of water, which is given off only at a temperature above 140 degs.

Aurine, which has been repeatedly crystallized from acetic acid or alcohol, does not melt at 220 degs.; at this temperature the crystals assume a darker shade, which disappears again on cooling, without any appearance of alteration in the substance. When more strongly heated it melts, emitting at the same time the odor of phenol, and solidifies again, on cooling, to an amorphous, beetle-green mass. Aurine dissolves readily in alkalies with a magenta-red color, and is precipitated from this solution by acids as a crystalline powder.

Aurine, crystallized from a mixture of alcohol and acetic acid, forms dark red crystals, moderately thick in comparison to length.

A variety of substances derived from aurine when treated with other bodies are given fully described by the authors.

When aurine is heated carefully in a combustion-tube, a reddish-colored oily liquid distils, and a large quantity of porus carbon is formed. The distillate is almost completely soluble in caustic potash, only a trace of a solid having the odor of diphenyl being left behind. The alkaline solution was decomposed with hydrochloric acid, and the oil dried over calcium chloride. On distilling it, a small quantity of water first passed over, and the boiling point then rose rapidly to 184 degs., remaining constant until the last drop had distilled over; the distillate solidified to a mass of needle-shaped crystals, and consisted of pure phenol, no cresol being present, the formation of which might have been expected if this compound took part in the production of the color.

We may, therefore, for the present assume that the aurine contained in the commercial product is identical with that obtained from pure phenol. It is known how easily one may be deceived by the apparent purity of crystallized coloring matters; only a short time ago, Wichelhaus has again called attention to this point in his beautiful researches on the oxidation of phenol.

When aurine is heated with aqueous ammonia to 140-150 degs., a new coloring is formed, dyeing on wool and silk a redder shade than aurine, and occurring in commerce under the name of "red coraline," or "pæonine."

Another derivative of aurine, called "azurine" or "azuline," is produced by treating it with aniline. An examination of this blue compound has so far yielded the following results.

When aurine is gently boiled with aniline and a little acetic acid, the solution soon assumes a pure blue color. On boiling the product with dilute hydrochloric acid, in order to remove an excess of aniline, a blue resinous substance is obtained, consisting of a mixture of different bodies' which are partly soluble in alcohol and acetic acid, and partly insoluble in them.

By heating the above mixture on a water-bath, a blue solution is formed in 16-20 hours, which, however, also contains several bodies. A portion of the product is readily soluble in caustic soda with a purple color, and preci itated by acid from this solution in blue flakes, which dissolve in alcohol and acetic acid. The portion which is insoluble in alkalies dissolves completely in acetic acid and alcohol, with a fine blue color, but ether takes up only a part of it, forming a dark red solution, which on evaporation leaves a blue resinous body behind. The portion not dissolving in ether forms a dark blue powder with a golden reflection.

PARAFFIN FOR STOPPERS AND LABELS.

PROFESSOR MARKOE states that the practice in Boston is, not to varnish a label for acid bottles, but to use paraffin instead. They had applied it to a large number of bottles in the college laboratory, and it answered perfectly. The only thing necessary was to brush the paraffin on as hot as possible, so as to get a thin even coating; it looked as well as varnish, and stood a great deal better. It saved a good deal of trouble in sizing and varnishing, and five minutes after the bottle had been brushed it was ready for use. It has been stated that the use of paraffin could be extended a great deal further; that instead of sealing the tops of bottles—sample bottles of bleaching-powder, and for other purposes—it was very convenient to have a small porcelain dish with paraffin always ready, which could be placed upon a lamp, and as soon as it was warm to dip the top of the bottle in it, and that gave as good a sealing as sealing-wax, or better, and caused very much less trouble. It had also been proposed to use stoppers made of solid paraffin for soda samples; but he did not like this, because they broke so easily. What Dr. Lunge had found to answer perfectly well was to rub some heated paraffin upon the stoppers in place of tallow. He found it a great deal cleaner, and answering in every way for this purpose.

A RESINOUS EXPLOSIVE.

IT is stated by Alfred Nobel, of Hamburg, that a powerful explosive that may be safely handled, and one which will burn without exploding if unconfined, may be made by mixing nitrate of soda and resin, or their equivalents, with or without a little sulphur, with sufficient nitro-glycerine to form a stiff paste.

BLUE DYE DERIVED FROM CARBOLIC ACID.

THE carbolic acid is mixed with eight or ten parts stannate of soda. to which is immediately added concentrated sulphuric or hydrochloric acid. By using the former, a yellow substance is obtained, soluble in tartrate of soda and in the alkalies. By the addition of a large proportion of acid, the mixture becomes reddish brown, and all the carbolic acid dissolves. Combined with a large amount of water, the solution becomes red, and brown flakes, soluble in alcohol, are deposited.

These flakes give to alkalies a blue coloring matter, which is not precipitated either by water or alcohol. It has not been isolated, except on the fabric. The red watery solution treated with alkalies becomes green on account of the formation of the blue dye and a yellow substance. If a fabric of oiled cotton be plunged therein, it is rapidly colored orange, which tint, like the liquid, passes to green when acted upon with alkalies; but if the dyed material be finally left to the action of water, it becomes a sky-blue, which is almost unalterable by chlorine and the hypochlorites.

RECOVERING FAT AND COLOR FROM WASTE WASH LIQUORS.

A PROCESS for the above, suggested by Messrs. Thom & Stenhouse, of London, consists in treating the waste soap-liquor with a solution of muriate of lime and adding milk of lime until free lime remains in the mixture. After mixing thoroughly and settling, the supernatant liquor is drawn off. The precipitate, containing the fatty and coloring matter, is then treated with sufficient muriatic acid to decompose the fatty matter, but not the coloring; the whole is then strained through flannel, and the fatty and coloring matter left on the strainer are heated to melt and agglutinate the colored fatty substance, then cooled and pressed in bags to remove any watery solutions left by the first straining. The substance removed from the bag may be further heated to remove any remaining water, and the color combined with the fat may be separated by heat and pressure, or by treatment with hydro-carbons as a solvent.

DYEING OF FELTED FABRICS.

IN making feltered fabrics of a mixture of animal and vegetable fibres, it is found difficult to dye them evenly, as the vegetable fibre does not take the dye equally with the remainder. To overcome this difficulty, J. T. Waring, of Yonkers, N. Y., proposes to neutralize the vegetable matter by subjecting the felted fabric to an acid bath of from 6 deg. to 12 deg. Beaume, and then washing to remove the acid, after which, it is stated, the fabric will dye an even tint.

ANTHRAPURPURINE—A NEW COLOR.

BY W. H. PERKIN, F.R.S.—Commercial artificial alizarine, when introduced as a dyeing agent, was generally supposed to contain purpurine, owing to the pureness of the red colors it produced with alumina mordants; but in a paper which I had the honor of reading before the Chemical Society some time since, this idea was shown to be incorrect; in the same paper, however, the existence in this product of a coloring matter differing from alizarine was pointed out.

To obtain this substance, which I propose to call *anthrapurpurine*, from commercial artificial alizarine, I have tried various methods, amongst these repeated crystallization from solvents; but this has not enabled me to separate it perfectly from the alizarine and other products with which it is associated, although its solubility differs considerably from them. I was, therefore, obliged to have recourse to chemical processes for its separation.

Instead, however, of converting the commercial alizarine into a lake as I previously did, and then treating it with an alkaline carbonate, I find it more convenient to dissolve the crude coloring matter in dilute sodium carbonate, and then well agitate the resulting solution with freshly precipitated alumina, which combines with the alizarine, leaving the anthrapurpurine in solution. This is filtered off from the alizarine lake, heated to boiling, and acidified with hydrochloric acid. The coloring matter which is precipitated is then collected on a filter, washed and dried. The anthrapurpurine thus obtained is very impure. These impurities can be removed to a considerable extent by repeatedly boiling the product with alcohol, anthrapurpurine being but little soluble in that menstruum. I have usually performed this operation nine or ten times; but the residual product, after crystallization from glacial acetic acid, has not given very satisfactory results, although analyzed several times. To further purify it, I have found it best to digest it with boiling alcoholic soda, and collect the difficultly soluble sodium compound which forms on a filter and wash it several times with dilute alcoholic soda. This is then dissolved in water, boiled, and the coloring matter precipitated with barium chloride; the purple barium compound thus obtained is collected on a filter, washed a few times with hot water, and then decomposed by boiling with sodium carbonate; the resulting purple solution is filtered off, and the anthrapurpurine precipitated with hydrochloric acid. After this has been collected on a filter, it is well washed with water, dried, and finally twice crystallized from glacial acetic acid.

Anthrapurpurine, when heated, at first fuses and then evolves orange-colored vapors, which condense as yellowish-red leaves or needles, but by far the largest quantity of the substance is carbonated. It is difficultly soluble in alcohol and ether, but rather more soluble in glacial acetic acid.

It is deposited from the boiling acetic solution on standing in small fun-
goid-looking groups of minute orange-colored needles. These groups are
generally not more than 2 or 3 mm. in diameter, and from the direction of
the crystals generally appear lighter on the under side than on the upper.
They can only be seen to advantage under the microscope. As this substance dissolves but slowly in boiling glacial acetic acid, it is sometimes
necessary to distill off part of the acid before the resulting solution can be
made to deposit crystals.

Anthrapurpurine is very slightly soluble in water, and may be removed
from its aqueous solution by means of ether. When heated with powdered zinc, it yields a hydrocarbon in small quantities, which, when purified, has the fusing-point and other properties of ordinary anthracine.

Anthrapurpurine, heated under pressure with acetic anhydride in
excess to a temperature of 150 deg. to 160 deg. for four or five hours,
entirely dissolves, and the solution, on cooling, deposits a large quantity
of pale yellow scales; these are easily purified by draining off the excess
of acetic anhydride, and recrystallizing three or four times from glacial
acetic acid. • For the preparation of this compound, I find it is not necessary to use pure anthrapurpurine, the crude product before treatment
with alcoholic caustic soda answering very well; but in this case the new
compound should be recrystallized until the mother-liquors are of a clear
pale yellow color.

This substance is a triacetylanthrapurpurine.

Triacetylanthrapurpurine melts at 220 deg. to 222 deg. It is not very
soluble in alcohol, but is moderately so in glacial acetic acid. It crystallizes from this latter solvent in beautiful pale yellow glistening scales, as
already mentioned; it decomposes when heated with alkalies.

When added gradually and in small quantities at a time to nitric acid,
sp. gr. 1.5, triacetylanthrapurpurine dissolves without effervescence, forming a dark yellow solution; this, on being slowly added to a large quantity
of cold water, deposits a pale brown precipitate, which, when collected and
washed, dissolves in potash with a reddish purple color. This solution
does not appear to give any bands when viewed by the spectroscope. but a
considerable absorption in the orange and violet. The addition of acid
causes the coloring matter to separate as an orange precipitate. This substance dyes alumina mordants of an orange color, and weak iron ones of a
reddish purple.

On boiling anthrapurpurine with benzoyl chloride, hydrochloric acid
is evolved, and the coloring matter quickly dissolves. The resulting product when cold becomes a viscid mass, and is purified first by repeated
boilings with water, to decompose the excess of benzoyl chloride and
remove a large quantity of the resulting benzoic acid, and then by crystallization from glacial acetic acid. This latter operation requires to be
repeated, and, after a portion of the substance which at first separates as a

yellow powder has been filtered off, the mother-liquor, on being allowed to stand, gradually yields small groups of crystals of a dark yellow or brownish color in the form of rosettes. These, when viewed under the microscope, appear well formed and transparent.

The substance is tribenzoylanthrapurpurine.

Tribenzoylanthrapurpurine fuses at 183 deg. to 185 deg. It is moderately soluble in boiling glacial acetic acid. Alcoholic potash decomposes it.

Anthrapurpurine forms metallic derivatives, most of which are insoluble or nearly so in water, except those of the alkali metals.

Solutions of potassic and sodic hydrate dissolve anthrapurpurine, producing beautiful violet-colored liquids, which become blue in shade when heated. They are not so blue as those obtained with alizarine. The sodium derivative is difficultly soluble in alcohol. Anthrapurpurine also dissolves in solutions of the alkaline carbonates with a reddish purple color, from which it is precipitated by carbonic acid. A solution of sodic bicarbonate, if gently heated, dissolves anthrapurpurine more readily than alizarine.

With the chlorides of calcium, magnesium, barium, and strontium, its ammoniacal solution gives purple precipitates. Specimens of the barium precipitate have been frequently examined after drying at 170 deg. to 180 deg., but the results have not been satisfactory, the percentage of barium being about two lower than that required by the formula. With alumina it forms a red lake somewhat similar to that produced with purpurine.

Cupric acetate when added to an alcoholic solution of anthrapurpurine, changes it from yellow to a beautiful purple, which cannot be distinguished from that produced with alizarine under the same circumstances.

When an ammoniacal solution of anthrapurpurine is heated to 100 deg. in a sealed tube for some hours, its purple color changes to indigo-blue. This solution, when acidified with hydrochloric acid, deposits the new product as a dark purple precipitate, which dissolves in ammonia and in sodium carbonate with a blue color, but in caustic alkalies with a red purple color. It dyes alumina mordants purple, and weak iron mordants an indigo-blue

Anthrapurpurine has about the same affinity for mordants as alizarine. The colors it produces are also analogous to some extent, as it produces red with alumina, purple and black with iron mordants.

There is, however, a considerable difference in the shade of color produced, the reds being much purer and less blue than those of alizarine, whilst the purples are bluer and the blacks more intense. The fastness of the colors against soap and light is quite equal to those produced with alizarine.

When used to dye Turkey-red, it produces very brilliant colors of a scarlet shade, which are of remarkable permanence.

COLORS ON TEXTILE FABRICS.

By L. Gabba.—The author remarks that, although dyers can occasionally guess the nature of the coloring matter employed in dyeing a fabric, yet they are very likely to err, so that it would be much better to have a methodical process for determining the nature of the coloring matters, similar to that employed in inorganic analysis. Insoluble aniline blue is changed by soda to a brownish red, whilst the soluble blue becomes colorless; azuline under similar circumstances changes to violet. In mixed vat and logwood blues, acids (other than nitric acid) destroy the latter and leave the former unchanged. When mixtures of Saxony and Prussian blue are treated with hypochlorites, the former is destroyed and the latter left. A mixture of indigo and Prussian blue acquires a greenish shade when treated with soda, owing to the Prussian blue changing to yellow. Molybdenum blue is recognized by the ash of the fabric containing molybdic acid and tin whilst ultramarine becomes colorless and evolves sulphureted hydrogen when treated with an acid.

DYEING FABRICS.

A new dyeing compound proposed by F. G. Graupner, of Evansville, Ind., consists of a base formed by dissolving and combining thirteen pounds of anvil dust (oxyduloid of iron) with twenty-four pounds of muriatic acid, evaporating the mixture to half its bulk, cooling, settling, and decanting the clear liquid. This base is added to quercitron water and extract of logwood, and the liquor boiled for fifteen minutes. The material to be dyed should be immersed in the boiling liquor for about twenty seconds, then wrung out and immersed in a vat containing about five hundred pounds of warm water, in which has been dissolved one pound of bichromate of potash, after which it is ready for starching and finishing.

IMPROVED PENCIL MACHINERY.

Orestes Cleveland, of the Dixon Crucible Company, of Jersey City, has introduced new machinery for the production of fine lead pencils, and the Company has been working very successfully for more than a year, turning the finest grades and styles almost without hand labor. The machines take the cedar, already sawed to size, plane it, cut the grooves for the leads, shape the pencils, apply the color, varnish and polish them with a rapidity and perfection that would surprise the hand labor heretofore employed for such work.

The varnishing machine, for instance, feeds itself from a pile of pencils, and turns them out, dried and finished at the rate of 106 per minute, with no attendant but a little girl 12 years old.

W.G.Jackman.

CHARLES H. LEONARD.

CHARLES H. LEONARD, second child of George and Cynthia Leonard, was born in Middleborough, Plymouth County, Massachusetts, September 23rd, 1813. Here he passed the first seven years of his life, and then removed with his parents to the neighboring town of Rochester. At the age of thirteen we find him again in Middleborough, attending school at the academy in that place. Long years of study were not then the privilege of many boys in the country towns in which his earliest years were spent, and he, too, at the early age of sixteen, left school and home to seek his fortune. Already there were developed in his character many of these noble traits that afterwards distinguished him as a man. One who knew him best, writes : " He was always the noble, generous boy, and the universal favorite of his companions; the younger and weaker found in him a sure defender, and parents and sisters a dutiful, helpful son and brother." As in the case of the great majority of our successful merchants, his earliest experience of business was in the disciplinary school of clerkship, the first year of which was spent in the employ of a Mr. Beaurasso, of Wareham, Massachusetts, and the three or four succeeding years as a clerk of Mr. Alfred Gibbs, a commission merchant of New Bedford. And then another and a bolder step was taken. He, as so many other sons of New England have done, caught the fever of Western migration, and for more than a year had personal experience of the opportunities and disadvantages of a pioneer life. Not satisfied with the inducements which the West held out to him, he returned to New Bedford; and it is at this point that his career as a merchant begins to open.

Not long after his return, an uncle, confiding in his integrity and skill, entrusted him with a quantity of oil to take to New York and sell. Day after day passed and the oil remained unsold, but through no fault of his, we may be sure. The market continued unfavorable; and regarding his undertaking as a defeated venture, he finally determined to take back his cargo to New Bedford, and fixed the day for his departure. But at the time appointed head winds set in; and the vessel by this circumstance was detained two days. Meantime his efforts to find a market were not abated, and within the period he was able to effect a most advantageous sale, which so elated and encouraged him that from that moment he determined to try his fortune in New York; and he was often heard to say that "his fortune as a business man was decided by a change of wind." In 1838 Mr. LEONARD started in the oil business in Front street, near Rosevelt street, New York. In 1840 he and a cousin (Mr. Horatio Leonard) formed a business connection at number 140 Front street, and during that year established an oil manufactory at Brooklyn; but the undertaking was not a successful one, and at the end of three years the firm failed and was dissolved. A settlement with creditors was before long effected, and then Mr. LEONARD started alone in the same place as a manufacturer and merchant in the sperm and whale oil trade; and as year succeeded year his business gradually enlarged, his connections increased in number and importance, and his success became more and more assured, until at length he stood among the foremost in his branch of trade in the United States, which position he maintained for a number of years until his death.

And here, in view of delinquencies too prevalent then and now among merchants of all classes, it is proper and pleasant to refer to a circumstance connected with his first unfortunate venture as a merchant in New York, to which reference has just been made.

A voluntary settlement with creditors was effected soon after the failure, as we have said, by which he was enabled almost immediately to resume business, and, not long after, upon certain partial payments being made, to obtain a complete discharge of all liabilities. But Mr. LEONARD's sensitive honor would not suffer him to forget his early obligations, and in 1865 he astonished his former indulgent creditors by paying to them the whole balance of the old claims (which they themselves had canceled and forgotten), together with more than twenty-two years' interest.

This incident is related chiefly for those who had no personal knowledge of Mr. LEONARD. To intimate friends and acquaintances it would have seemed an anomaly had it not occurred, so closely associated with his honored name, in all their minds, was the idea of strictest fidelity.

On October 13th, 1845, he married Elizabeth E., adopted daughter of Captain Robert and Anna Gibbs of New Bedford—a union which proved uncommonly felicitous, and to which by reason of its sympathies, its happy influence and encouragements, must be attributed no small part of the noble results of his life.

Mr. LEONARD's career as a New York merchant was embraced in the period from 1838 to 1868; and within those thirty years he achieved a marked success in business, and a reputation for integrity and honor, which must have been far dearer to him than his wealth.

His principal field for business operations was, of course, the city of New York; but, after the first few years of initiative struggle, and when increase of business admitted and called for an increase of facilities, he established in New Bedford a factory for the manufacture of oil and candles, and there founded a most important agency.

Probably no one enjoyed the confidence of business associates to a greater degree than Mr. LEONARD; and his responsibility and reputation becoming established, solicitations without number were made to draw him into outside speculations and ventures, or to induce him to lend his support to undertakings that presented no small array of reputable names. But he was unswerving in the course which his best judgment had marked out for him, which was to concentrate his energies upon his own legitimate business, and he refused all solicitations to deviate from it. In the earliest period of the petroleum excitement the most tempting offers were made to draw him into the ring of speculators, and immense sums were almost ready to be pledged to him if he would only lend his name to the schemers' plans. But he was true to his conscience and his rule, and no bubble was by his countenance or assistance set afloat upon the world to burst in the hands of deluded purchasers.

In all his dealings he was integrity itself. Every contract written or verbal was fulfilled not only in its letter but its spirit, and that, too, when his business operations had become very extensive, embracing many heavy contracts, which involved in their execution many intricacies of detail, and which to a man of weaker virtue would have presented many temptations to artifice and fraud—as witness his dealings with the government.

For a number of years he was awarded the contract to furnish oil to the United States Light House Board, which in every instance was executed with the highest approval on the part of official inspectors; and no whisper of suspicion was ever heard that the government was at any time defrauded of a scintilla of its rights—a circumstance to be noticed in these latter days.

Mr. LEONARD was of such strict integrity that no one could have had dealings with him without discovering this trait of character. No detail was too insignificant to engage his attention when another person's interest was involved in it. He was conscientious in matters great and small. He held strict rule over himself in all his affairs, and was exceedingly sensitive to every call of duty, so that in dealing with him no one needed to be on guard to protect himself against even the slightest injustice or wrong. Had a mistake occurred, which it was Mr. LEONARD's duty to rectify, it would have been done at once, and without solicitation, cost what it might. Being thus strict with himself he expected fair dealings in others. There was a shrinking of his whole soul from any contact with a bad or untruthful person, though he was the last one in the world to harbor resentment.

No better testimony to a man's worth could be obtained than that of former employés, before whom he went in and out for many years, under the varying circumstances and conditions of active business life, and to whom ungenerous traits, if any there were, though masked and concealed from the outside world, would surely in time be revealed. Judged by this standard, Mr. LEONARD's character appears to have been of singular excellence. From confidential clerk to humblest servant, all speak of him only in terms of reverence and affection. His old porter who was in his employ for nearly thirty years, and who was remembered by him in his will, would gladly spend hours in dilating upon the single subject of Mr. LEONARD's goodness.

Mr. LEONARD was a man of excellent judgment, and in the management of business was sagacious and prudent, never yielding to the speculation mania of the day, but directing his course according to approved, conservative principles. He knew his business in all its details ; and he was attentive to all the signs and movements of the day affecting his trade. While watchful of his own interests, he was equally mindful of his neighbor's, and would rather suffer loss than do wrong. And there was such dignity and firmness in his rectitude, that he rarely had to repel the aggressions of others.

He was never harsh in insisting upon his strict legal rights ; and many a brother in the trade, we are told, less sagacious in reading the signs of the market, and less able to bear the consequences of a mistake, has been wholly released from the burden of his contract, or helped to execute it by Mr. LEONARD's yielding up half the profits of the bargain. His early struggles for success had left a deep impression on his mind, and he knew how to sympathise with all who were in such a case. He could not see a young man in conflict with hard fortune but the old days would at once come back again, and his whole heart would go out to him in tender sympathy; and when opportunity offered, he was always ready with encouraging words and a helping hand.

He was also a man of great charity of spirit, and distributed his bounties with a most generous hand; but the best of his benefactions were such as no society or corporate body could take notice of, and are only recorded in the grateful hearts of the needy and deserving whom they reached and blessed.

By profession of faith Mr. LEONARD was a member of the Presbyterian Church, having united with the Mercer street church, New York, when under the pastoral charge of the Rev. George L. Prentiss, D.D. He was afterwards one of the most active and munificent of the founders of the Church of the Covenant of the same city, of which Dr. Prentiss became pastor; was elected one of its first trustees, and remained such until his death.

In speaking of Mr. LEONARD in connection with his business, we have said that it was his rule and practice to concentrate his energies upon that which came legitimately within its compass; but when we come to consider him as a member of society we find that his sympathies were not limited, but were as broad as society itself. There was no important question of the day, whether of church, State, or society, in which he did not take a deep personal interest.

But it is not as a merchant, or a member of society in respect to public questions, that those who knew him best in life now love most to dwell upon. It is the liberal-minded, unselfish, sympathetic, tender-hearted, generous and just man that we picture to our minds; he, who could not only endure but sympathise with another in an opinion not his own; he, who was always ready with help when help was needed, and with such delicacy of ministration that its value was found as much in the manner as in the substance of it; he, whose sympathy could detect a sorrow before it had uttered a sigh, and the want that had never been confessed; whose friendship was constant through sunshine and cloud;

" Who spake no slander; no, nor listened to it."

Mr. LEONARD's life and activities were not confined to the city of New York; and they who knew him simply as the upright and successful merchant of Front street, or the respected member of New York society, will have had only a limited view of his excellent character, and that in which his most amiable qualities were not displayed.

It was the writer's privilege to know him intimately in New York, and also in the quiet town in Massachusetts where he had his country residence, and was accustomed to spend a considerable part of every year. It was in this place, probably, that his true character was best exhibited; and no one's memory could be more lovingly and sacredly cherished than is Mr. LEONARD's to-day in Rochester.

Here his benevolence reached every person and every institution. He was the benefactor of the church; the academy was remodeled and rebuilt at his expense, and materially assisted in its support; and every worthy object and enterprise was sure of his encouragement and aid. And not only was his benevolence shown in response to others suggestions, but what is rarer to find, he was almost always the first to discover the want which his charity supplied or relieved. Besides this, his whole influence was so salutary through all that community, that, by precept and example, he moved and excited all others to greater kindliness of word and act.

So reluctant was he to disclose his charities that but a small part of his good works will ever be known to the world; and such as now appear are due only to the thankfulness of those whom his kindness reached.

The gratitude of the poor woman who, bed-ridden for years, was brought by him from Rochester to New York, placed under the care of skilful physicians, and there tended for months at his expense, would require much time and many words to express; but it would tell the story of only one of his many beneficiaries. As an example of his peculiar benevolence the writer remembers an incident that was related to him when on a visit in Rochester, and then referred to as illustrating Mr. LEONARD's general habit. There had been one of the severest of New England snow storms; and, while the roads were still unbroken, and the scattered houses had become almost isolated by the depth of the snow, Mr. LEONARD caused his horse to be harnessed, and, taking provisions in his carriage, forced his way through the drifts until he reached the house of a poor neighbor, nearly a mile distant, and there deposited his gifts. He had not heard that the family were then in actual want, but he knew their limited means, and the thought of their possible necessities would not suffer him to rest until this deed of mercy was accomplished.

And thus example might be added to example, almost without end, illustrating the different traits of this truly good man; but the limit allowed us for this sketch has been passed, and we must now leave him to the love that cherishes his good name and to his works which follow him.

Mr. LEONARD died in Rochester October 24th, 1868, and there is his grave; and his memory, as of a just man,

"Smells sweet and blossoms in the dust."

The following letter from Mr. LEONARD's old pastor and friend, the Rev. Dr. Prentiss, now Professor in the Union Theological Seminary in this city, will be here in place:

NEW YORK, Jan. 21, 1874,

GEORGE B. BONNEY, Esq.

MY DEAR SIR: I have read your sketch of the life and character of the late Mr. CHARLES H. LEONARD with great interest and satisfaction. It is a truthful picture, rather under than over-drawn, of one of the worthiest men I have ever known. How many pleasant memori's—not of him only, but of New Bedford and Rochester, and of old friends now gone with him to the better country, its perusal revived! The names of some of these friends are in my mind indissolubly associated with his. I cannot think of him, for example, as I knew him in New Bedford nearly thirty years ago, without at the same time thinking of Captain Robert Gibbs, that whole-souled, noble man, and of Mrs. Gibbs, that most kind-hearted and excellent woman, under whose roof he found the greatest earthly treasure and comfort of his life.

My more intimate acquaintance with Mr. LEONARD began at his marriage and it ended only at his grave. From the time of my settlement in New York in the spring of 1851, he was my parishioner as well as my friend; and I had constant occasion to see him in all the varied and changing phases of life. And I can say with truth, that I never saw him when he appeared in any way otherwise than as the quiet, modest, fair-minded, upright, amiable, high-toned gentleman. He united with the church under my charge in 1857, I think; and his religious character, as it unfolded itself, was marked by the same attractive and solid traits which distinguished him as a man. His piety was not demonstrative, it was rather of a shrinking and reticent temper; but it gave ample roof of its sincerity and power by the benign and excellent fruits that adorned its path. In all the movement which issued in the establishment of the Church of the Covenant and in the erection of its beautiful sanctuary, chapel and parsonage, Mr. LEONARD's counsels and influence were invaluable to me. He was one of the Trustees, served on one of the most important building committees, and was himself, again and again, a liberal contributor to the object. I have always regarded him as one of the founders of the Church of the Covenant.

I shall not soon forget the day of his burial at Rochester. It is not very often that a country village witnesses such a scene, or follows to his grave one so beloved and esteemed as a friend, a neighbor, and a benefactor. The venerable fathers and mothers of that honored old town were there; the strong men, who had grown up with him from boyhood, were there; the young men and maidens were there; the poor were there; the school and the academy and the chúich weie theie; iepiesentatives of the adventurous enterprise and wealth and social cultuie of New Bedfoid weie theie; and nobody was there who did not give evidence of the heart-felt respect and affection with which Mr. LEONARD had been universaily regarded, and of the deep unfeigned sorrow caused by his death.

Believe me, my deai sir,

Most tiuly youis,

GEORGE L. PRENTISS.

PETROLEUM.

Its Origin, Description, and History.

ANTIQUITY OF PETROLEUM.

DEUTERONOMY XXXII, 13. And he made him to suck honey out of the rock, and oil out of the flinty rock.

JOB XXIX. 6. And the rock poured me out rivers of oil.

MICAH VI, 7. Will the Lord be pleased with thousands of rams, or with ten thousands of rivers of oil.

Evidences of petroleum are to be found among the ruins of Nineveh, whose existence dates back more than two hundred years before the Christian era.

Plutarch describes the spectacle of a sea on fire, or lake of inflamed petroleum, near Ecbatana.

There is a tradition in Venango County that the oil springs on Oil Creek formed a part of the religious ceremony of the Seneca Indians, who formerly lived on these wild hills.

The aborigines dipped it from their wells and mixed it with their war paint, which is said to have given them a hideous appearance, varnishing their faces, as it were, and enabling them to retain the paint for a long time. to keep their skin entirely impervious to water.

The use of this oil for their religious worship is spoken of by the French commander of Fort Duquesne in a letter to General Montcalm:

" I would desire to assure you that this is a most delightful land. Some of the most astonishing natural wonders have been discovered by our people. While ascending the Alleghany River, fifteen leagues below the mouth of the Conewango, and three above Venango, we were invited by the Chief of the Senecas to attend a religious ceremony of his tribe. We landed and drew up our canoes on a point where a small stream entered the river.* The tribe appeared unusually solemn. We marched up the stream about half a league, where the company, a large band of Indians, it appeared, had arrived some days before us. Gigantic hills begirt us on every side. The scene was really sublime. The great chief then recited the conquests and heroism of their ancestors. The surface of the stream was covered with a thick scum. which burst into a complete conflagration. The oil had

---

* The small stream spoken of was evidently Oil Creek, and that upon marching half a league above that stream, they have probably reached Rouseville, where the Cherry Run flows into the stream, and where the largest oil wells have since been found.—ED.

been gathered, and lighted with a torch. At the sight of the flames, the Indians gave forth the triumphant shout that made the hills and valleys re-echo again. Here, then, is revived the ancient fire-worship of the East; here, then, are the children of the sun."

Under different names petroleum has been known for more than two thousand years, having been found in widely distant sections of the earth. The generic name by which we designate it is a Latin word compounded of *pctra*, rock, and *olcum*, oil this name being given because of the fact that the fluid issues from rocky formations. Its natural composition is not fully determined though it consists chiefly of oily hydro-carbons which hold in solution paraffine and bitumen, or asphaltum. In some scientific works fluid petroleum is described under the name of "naphtha" oil, while that which is less fluid and more impregnated with asphaltum is called "bituminous" oil. Petroleum has become one of the most valuable of the world's products, and all its correlative parts, after refining, have a market value. Nothing except gas has greater illuminating power than the white oils. The gas from crude petroleum and from gas wells in the subterranean neighborhoods of petroleum deposits has been utilized for the lighting of buildings and streets. Naphtha is one of the best and cheapest substances for cleansing woolen goods; and even by a mere mechanical process, without chemical combination. petroleum may be converted into a gas available for burning under certain conditions; and the solid residuum left after the first refining process is finally converted into paraffine. Benzine a fine product of naphtha, has also been used as a substitute for turpentine. Petroleum has been used in toilet soap; as a substitute for fish oil in tanning; and as a medicine both internally and externally; and those who have given its properties the most philosophic consideration profess to believe that uses entirely novel and profitable will yet be found for it.

Petroleum is mentioned by both the Greeks and the Romans. The latter called it "bitumen." In Zante, an island of the Ionian group, celebrated chiefly for its immense production of currants. there is an oil spring still flowing which was mentioned by Herodotus more than 2300 years ago. The ancient Sicilians used it in their lamps in preference to fish oil. The streets of Genoa and Parma, Italy, have been lighted with it for two centuries. At Baku, in the Caucassian province on the Caspian Sea, there are extraordinary manifestations of petroleum and gas, extending over an area twenty-five miles long and half a mile wide. The geological formation there is of a porous argillaceous sandstone belonging to the tertiary period. Oil from this belt has been immemorially used in Persia for both the common and the sacred fires. The Burmese Empire, in India, has been for centuries supplied with petroleum from the Rangoon district on the Irrawaddy river, which yields annually 400,000 hogsheads from 520 wells. It is there used as a medicine, as a burning oil, and to preserve timber from the attacks of insects; this oil is about the same color as the Pennsylvania oil, but of heavier specific gravity. Off the Cape de Verde Islands a species of petroleum may be seen floating on the ocean; and to the south of Vesuvius a spring of it rises through the Mediterranean sea. Such is a summary of the principal facts concerning foreign petroleum, from which it will be seen that, instead of it having been introduced into trans-Atlantic countries from America, as many people believe, it was known and employed in various ways for many centuries before our country was discovered.

It is, however, in North America that the largest and most valuable deposits of petroleum have been found. These exist in Northwestern and Southwestern Pennsylvania, West Virginia, Ohio, Kentucky, New York, Canada, Kansas, California, and to a certain extent in Michigan, Indiana, Illinois and Iowa. Some geologists believe that petroleum originates in coal beds, while others assert that the coal is formed from the oil; but in view of the fact that the nearest coal beds in Pennsylvania are thirty miles distant from the source of the largest petroleum supply, and that no coal has ever been found in boring successful wells, it is entirely fair to assume that petroleum is not a coal oil. Another geological theory is that the oil has been distilled by some natural subterranean chemical agency from animal or vegetable matter, and that the supply must some day cease. Others, again, aver that, whether petroleum be the product of animal, vegetable or mineral matter, the process of distillation is contemporaneous, and not of a past period, and that the supply is inexhaustible. The last named theory is the most probable, and receives strong apparent confirmation from the fact that the wells of Baku and Rangoon are as productive now as they were a thousand years ago. Old wells have given out there, as here, but new ones are continuously found, and the aggregate product of the Asiatic districts does not diminish. This fact necessarily tends to allay the apprehensions of those who doubt the continuity of the supply of American petroleum.

Petroleum is sometimes traced to beds of lignite (that is, wood carbonized to a certain degree, but retaining its woody texture), and sometimes its source cannot be discovered. In the United States and Canada the sandstone formations are the most productive of oil. In regard to the " dip," or downward course, of the American oil-bearing strata, in Pennsylvania it is nearly southward, and in Ohio and West Virginia eastward of a southerly direction. It is in places which present appearances of upheaval, forming cracks and fissures in the rocks, that borers look most hopefully for oil in large quantities. It has been determined with approximate certainty that a "flowing" well of petroleum exists exclusively in cases where the oil-chamber has been struck before the gas-chamber in the earth. In this instance the oil flows spontaneously and sometimes with immense force until the gas-pressure is exhausted, when the oil must be pumped. If the gas-chamber is struck first the well never flows. When a flowing well is first struck, water almost always makes its appearance, in various proportions in different cases, until it has flowed for some days. The highest yield of a flowing well was 5,000 barrels per day, but this enormous product continued only for a few days. A few other wells have yielded between 2,000 and 3,000 barrels per day for a short time, when the gas-pressure subsided, and from flowing wells they became pump wells. The flowing wells run the lighter grades. The heavy oils require to be pumped.

The oil of different districts varies materially in specific gravity, and consequently in value. The lighter oils are more valuable for making burning oil, provided they are not too light, and the heavier for lubricating purposes. The Pennsylvania petroleum, at the present time, runs between 44 and 50 degrees specific gravity by standard hydrometer. The average of the West Virginia oil is about 38, and is used for lubricating. The heaviest terrene oil ever known was found at Mecca, Ohio. Its density was between 27 and 28 degrees, and none other ever equaled it as a lubricator, especially for heavy machinery. The wells there, however, virtually gave out some years ago. The refiners of the best oil for illuminating purposes prefer crude oil of 44 degrees gravity, or somewhat lower than that, because gravities of oil for burning that scale upward to 44 degrees usually yield those percentages of lamp oil, naphtha and residuum that most nearly meet the demands of the markets for the three products of petroleum. In regard to the depth at which oil is produced, it has been found, during twelve years of experience in our country, to vary between 100 and 1,300 feet. The "third sand rock" of the Pennsylvania geological series, in which the larger deposits of light oil are found, ranges at a depth of between 300 and 1,200 feet. The majority of productive wells there have been between 400 and 600 feet deep. The lubricating oils have commonly been found at depths varying between 70 and 180 feet.

The existence of petroleum in the neighborhood of Oil Creek, Pennsylvania, was undoubtedly known to a race superior and anterior to the Indians. In several parts of the Alleghany Valley the early settlers found pits of about twenty feet in depth and between six and eight feet in diameter, carefully walled around with timber which the petrolized waters had preserved from decay, and in which were found notched logs which served as ladders. The Indians could give no account of these pits other than that they must have been dug by an earlier race. Oil is still found in these pits. In druggists' stores it was sold as a cure for rheumatism. and bore the name "Seneca Oil."

In the year 1845 Mr. Lewis Peterson, Sr., of Tarentum, Alleghany

County, Penn., brought to the Hope Cotton Factory, at Pittsburgh, a sample bottle of what is now called petroleum. It came up with the water from his salt well, and was the source of much trouble therein. Mr. Morrison Foster, latterly of Cleveland, in conjunction with the manager of the spinning department of the mill, Mr. David Anderson, experimented with the oil, and soon learned that it could be combined with sperm oil in such a way as to form a better lubricator for the finest cotton spindles than the best sperm oil, which alone could previously be used for that purpose. The mixture cost about seventy cents per gallon, while the sperm oil alone cost one dollar and thirty cents. The saving was so great that a contract was made whereby Mr. Peterson was to supply two barrels per week, and for ten years this oil continued to be used at the Hope factory unknown to any except the proprietors. This is believed to have been the first practical use to which petroleum was put in America. From 1850 until 1855 it was extensively used in Pittsburgh, under the name of "carbon oil," for burning.

It is now (1873) about twenty years since attention was directed toward petroleum with the view of developing it in sufficient quantities, and treating it in such a manner as to make a cheap and good light. A company—the first ever organized—was formed by Messrs. Eveleth and Bissel, of New York, under the name of the Pennsylvania Rock Oil Company, Professor Silliman being at its head. Their operations were confined to collecting the surface oil until, in 1858, Colonel E. L. Drake, of New Haven, Connecticut—to whom belongs, undisputedly, the distinction of having devised the means whereby petroleum became of such immense commercial importance—was engaged to visit the Oil Creek valley, where he set about sinking a well on Watson's Flats, a mile and a half below Titusville. This was the very first attempt ever made to drill a well for oil and it was a failure. But the practicability of boring a well through the hardest strata had been demonstrated, and that was a gain. Another attempt was made, and was a success. The drill probed an oil cavity at the depth of seventy-one feet, and, on the tools being withdrawn, the oil rose to within five inches of the surface. It was pumped off, and yielded at first four hundred, and afterwards a thousand gallons of oil per day. The result was the most intense excitement among the people of the valley, who immediately commenced sinking wells on their own account or leasing their lands to other parties who desired to sink them. A very small proportion of the wells were then successful, and the demand for oil was limited. Many people regarded petroleum as a mysterious and dangerous commodity, and the sale was relatively small. Still, several of the adventurers were making fair wages, when the discovery of flowing wells suddenly revolutionized matters. Pumping oil at the rate of five to twenty barrels per day was a discouraging process when, sometimes at not more than a hundred feet distant, oil was spontaneously running from another well at the rate of 500 to 2,000 barrels per day. The flowing wells glutted the market and reduced the price, at one time, to ten cents per barrel. Lessees of pump wells fled in despair, in many instances leaving their machinery behind them, and not stopping even to surrender their leases. Some of their abandoned wells have since been worked, and more would be but for the impossibility of reaching the lessees, and the consequent fear of undertaking operations from which, if successful, the legal controller of the ground might oust the operators.

The first flowing well ever struck was on the McElhinney or Funk farm. It was called the Funk well, and was struck in June, 1861. It began flowing 250 barrels per day, and maintained that average for fifteen months, when Mr. Funk, previously very poor, found himself very rich. The next and a contemporary spring was the Phillips well on the Tarr farm, a few miles from Titusville. It flowed 2,000 barrels daily. While the Phillips was at the hight of its productive power the Empire was discovered, not far from the Funk well. The Empire flowed 3,000 barrels per day. It is virtually exhausted now, but its maximum capacity has never been equalled by that of any other well. In these "early days" (as oil men call them) of flowing wells, the supply of oil so far exceeded the demand, and it was so difficult to convert the property into money, and as coopers would work only for cash payments, it was impossible to get anything like a sufficient number of barrels to put the commodity in a marketable condition. The Sherman was the next flowing well. It was put down after the greatest financial difficulties, in 1862. At last, after almost every means of borrowing money and selling stock were exhausted, oil was struck and flowed at the rate of 1,500 barrels per day, and continued at that figure for several months, when it declined to 700 barrels daily. It flowed altogether for twenty-three months, and then stopped. For the first year the low price of oil prevented the proprietors from making money, but during the last nine months they realized an immense fortune. It is now of very little value. In March, 1863, the Caldwell well was struck on the Foster farm, not far from the Sherman, and flowed 1,200 barrels daily. Two months afterward the well since called the Noble and Delamater, but then known as the Farrell, was found close to the Caldwell, and commenced flowing at the rate of 2,000 barrels per day. The fountain from this well was stronger than that from any other of which a record has been kept until the discovery of the "Cash-Up" well, October 21, 1871, which for some weeks exceeded that of the Noble and Delamater in both volume and sound—the sound resembling, in both instances, that of the escape of steam from a boiler through a three-inch pipe. From the year 1863 onward the discovery of a flowing well ceased to be considered extraordinary. No exact record of the spontaneous wells of large productive capacity has been kept, but the men oldest to the business say that between 250 and 300 have flowed from 100 to 3,000 barrels per day between the date of the first well (above given) and the present time.

In the early days of oil enterprise, and after the yield had become large, great difficulties existed in the way of getting the product to market. There was no available railroad transportation, and it had to be floated down Oil Creek to the Alleghany river and thence to Pittsburg. The supply of flat boats on the creek and river was far too small for the requirements of the trade. When boats could not be had the oil barrels were formed into a raft and lashed together. Scarcity of barrels frequently occurred in early times. When this was the case the flat-boats were made oil-tight and the oil poured into them in bulk. At the shallow places—and they are numerous—in the creek large dams were made, and at an appointed time a pond-freshet swept boats and rafts down to the river. This process of getting oil to a market was frequently very amusing, but the amusement was very expensive. Accidents would happen through the carelessness or lack of skill of certain boatmen, and when the dam was cut away, the whole mass of boats, rafts, tank-boats, etc., would sometimes be broken and stranded and a large amount of property destroyed. The burning of these boats was not an infrequent occurrence. On May 12, 1863, a very large number of tank-boats took fire on the creek above Oil City. The burning oil ran out on the rapid stream and set fire to everything combustible along its banks, and very nearly consumed the city itself. The Alleghany river and the creek at this date and for a few following days had a line of fire almost unbroken for twenty miles. The bridge at Franklin was totally consumed. Gas wells, flowing wells and others frequently took fire and were extinguished with great difficulty.

Certain places on the Alleghany river are lighted with gas from wells. One of the most weird and beautiful sights in the world is to witness these immense gas torches—"pillars of fire"—illuminating a vast area covered with snow. The effect is extremely impressive and sublime.

EARLY PRICES OF PETROLEUM.

The first market quotations of refined petroleum were given in the fall of 1860, and ranged from 70 to 75 cents per gallon, after which prices varied from 60 to 80 cents during the remainder of the year.

The first sales of crude oil were noticed in 1861, and were at from 20 to 25 cents, refined at the same time commanding 60 to 70 cents, closing that year with sales at 15 to 24 cents, as to gravity, for crude, and 37½ to 50 cents for refined, as to color. The above quotations were in gold.

In 1862 crude opened with sales at 16 to 20 cents and refined at 30 to 40 cents, gold. In May crude sold as low as 9½ and 11 cents, as to gravity, and refined 19 to 25 cents, gold, as to color and test, 110 to 120 degrees.

In September, 1862, the first sales of bonded refined petroleum were made at 31 cents, currency (gold 120¾), for white; heat test, 120 to 125 degrees.

In October, 1862, commenced the speculative fever, which culminated in the third week of November, when sales of crude were made as high as 55 cents, currency; refined, duty paid, $1 08, currency, per gallon (duty 10 cents), and 96 cents, currency, in bond; and naphtha $1, currency, per gallon (gold 130½), all including packages. In the beginning of October 1862, crude was quoted at 17 cents, currency; refined, duty paid, 28 to 35 cents, currency, and bonded refined at 30 to 32 cents, currency, and naphtha 23 to 30 cents, currency, per gallon (gold 123¾).

About the 20th of November, 1862, when almost everybody had invested in refined, and prices had attained their highest point, a sudden and violent collapse followed, involving speculators in heavy losses. The advance in November for ten days was 30 cents on crude, 41 cents on refined in bond, 45 to 48 cents on refined, duty paid, and 60 cents per gallon on naphtha. At this time several parcels of refined were taken out of ships and sold at a profit, which could not have been hoped for by its sale in Europe, and this fact contributed somewhat to the reaction which followed.

At the close of 1862 crude sold at 25 to 27½c; refined in bond 44 to 50 cents on the spot, and 40 to 42½c for future delivery; refined, duty paid, 45 to 50 cents, and naphtha 30 to 35 cents, currency, per gallon, packages included, and gold at 133⅓.

Crude petroleum at the wells sold as low as 10 cents, gold, per barrel of 40 gallons measure in 1861, and as high as $10 25, currency (gold 208¼), per barrel of 40 gallons measure, barrel $1 25 to $1 50, inclusive.

Crude petroleum at 10 cents for 40 gallons was, for want of barrels, let run into deep holes dug into the earth, and thus allowed to waste for weeks.

The foregoing is an accurate compendium of all the facts of interest relating to the early history of petroleum. From these we pass to the consideration of petroleum as a commercial property, the process of well-drilling and of refining, and the railroad and ocean transportation of oil, and the general statistics of the petroleum business.

COMMERCIAL VALUE OF PETROLEUM.

No absolutely accurate record exists from which to compile reliable statistics of the exact value of petroleum to our country; but a number of the oldest dealers, who have kept partial records of production and consumption since 1861, are of the opinion that the net profit which we have made on petroleum, and which we could not have made from any other of our products, is $100,000,000.

Fully 200 men and assistants are employed in our country as brokers in petroleum. The amount of brokerage varies according to the product sold. On crude oil it is three cents per barrel; on refined oil and naphtha, packed or in bulk, it is one-half of one per cent on value; on empty barrels three cents per barrel, and on residuum twelve and a half cents per barrel.

The brokerage is almost invariably paid by the sellers, and the total of brokerages earned has been estimated, upon competent authority, at about $1,000,000 per annum.

The consumption of petroleum in the United States is estimated at 8.000 barrels per day between December 1 and March 1, and at 6,000 barrels per day for the remainder of the year.

PORTS OF SHIPMENT—OCEAN TRANSPORTATION.

Philadelphia ranks next to New York as the principal port for the exportation of petroleum, but considerable quantities are shipped from Boston and Baltimore. More than ninety per cent, however, of the total export of petroleum is done on account of principals and agents located in New York, and this has been the fact for twelve years.

The exigencies of the ocean transportation of petroleum are so numerous and so peculiar as to constitute it almost a special branch of marine service, and the exportation of it for the last four years has been so large as to have aided materially in effecting a permanent increase on all freight rates. Two ships, specially fitted with tanks to carry crude oil in bulk, have been built and operated. One of them sailed hence for Antwerp with a cargo about three years ago, and was never heard from. The theory of her loss is that the gas from the oil found its way inside one of the signal lights at night (the danger of which is extreme, and necessitates in all vessels great precautions), and that the vessel was instantaneously blown to pieces. The other ship safely made two trips, but the enterprise of carrying crude oil to Europe in bulk was then abandoned because, as the vessel could bring no freight back, the undertaking was found to be too costly in competition with ships that could bring back miscellaneous cargoes as well as carry away those of petroleum.

A cargo of refined oil is comparatively safe, but one of crude oil or naphtha—particularly the latter—involves great and continuous peril on account of the extreme liability of contact between the gas which crude oil and naphtha emit and the necessary ship's lights, not to dwell on other shipboard fires.

BORING FOR OIL.

The derrick used for boring is a square frame of timbers, substantially bolted together, making an enclosure about forty feet high, and ten feet square at the base, tapering as it ascends. It is generally boarded up a portion of the distance to shelter the workmen. A grooved wheel or pulley hangs at the top and a windlass and crank are at the base. A short

distance from the derrick a small steam engine is fixed and roughly en-
closed. A pitman rod connects the crank of the engine with one end of a
large wooden walking beam, placed midway between the engine and the
derrick, the beam being pivoted on its centre, at about twelve feet from
the ground; a rope attached to the other end passes over the pulley, at
the top of the derrick, and terminates immediately over the intended bore.
A cast iron pipe, from four and a half to five inches in diameter, is driven
into the surface ground, length following length until the rock is reached.
(In the older wells the ground was dug out to the rock and a wooden
tube inserted.) The earth having been removed from the interior of the
pipe, the actual process of boring commences. Two huge links of iron,
called " jars," are attached to the end of the rope; at the end of the lower
link a long and heavy iron pipe is fixed, at the end of which is screwed a
drill, about three inches in diameter and a yard long. When all is ready
the drill and its heavy attachments are lowered into the tube and the drill
set in motion. With every elevation of the derrick end of the walking-
beam the drill strikes the rock, the heavy links of the "jars" sliding into
each other and thus preventing a jerking strain on the rope. The rock, as
it is pounded into a pulverized condition, mixes with the constantly drip-
ping water, forming a pasty mass. After a while the drill is hoisted and
a sand pump dropped into the hole. The sand pump consists of a copper
tube about five feet long, and is a little smaller than the drill, having a
valve at the bottom, opening upwards and inwards. As the tube is pushed
into the hole the pasty mass rushes into it through the valve and remains
there. When this has been done several times the tube is hoisted and
emptied, this operation being repeated until the hole is clear, when the
work of drilling recommences.

The drill is turned by the hand of the "borer" after each stroke,
by which means a nearly circular hole is produced from the first. A
" reamer " is then put down, which makes it perfectly round.

As the wells get down to points where the first indications of oil are
reached, the contents of the sand pumps are anxiously examined. The
oil borers have a geological system of their own, the prominent points
whereof are three layers of sandstone. The "first" lies immediately below the alluvial deposit. The "second" is at a variable depth of 100 to
300 feet, and here the primary indications of oil are commonly reached.
Some wells—especially those bored between six and ten years ago—go no
lower than the second sandstone, but the general plan is to proceed downward into the "third," where the most continuous deposit of oil is usually
found.

It frequently happens that the drill breaks and becomes fixed in the
hole. In such cases nothing can be done until the tool is removed. The
upper portion of the boring instrument is taken off, and a pair of nippers

or clamps let down to grip and extract the broken drill. Some men formerly made the extraction of tools a special business, and exhibited great ingenuity in their devices to overcome the difficulties. There are instances of wells having been abandoned on account of the impossibility of removing broken drills.

When the hole has been sunk far enough to strike oil, the next thing is to tube the well. An iron pipe, with a valve at the bottom like the lower valve of a pump, is run down the entire depth of the well, the necessary length being obtained by screwing the sections firmly together. If the oil does not flow spontaneously, a pump box attached to a wooden rod, also made of sections screwed into each other, is inserted in the tube, and the upper end of the rod is attached to the walking beam. In boring for oil, springs of water are, of course, cut through and the water falls into the hole; being heavier than the oil, it lies at the bottom, and would enter the pump tube except for the very ingenious contrivance known as the "seed bag," which is made of leather, resembling somewhat a bootleg in shape, filled with flaxseed, and crowded down to the proper place with the iron pipe, where the seed swells and forms a water-tight packing between the tube and the rock. At times the seed bag slips or bursts, when the well fills with water and the tube must be pulled up for necessary repair and re-arrangement. The quickest time ever made in well boring through hard rock was four inches in five minutes. It sometimes happens that after a well has been yielding for months it stops and will not produce another drop.

If such stoppage be not the result of the absolute exhaustion of the well, it is caused by the thickening of the paraffine matter usually at the bottom of the hole. In the latter instance nothing but the explosion of a torpedo at the locality of the obstruction is sure to restore its productive power.

The total cost of furnishing the apparatus for working an oil well is about 4,000 dollars. The items are as follows: An engine with boiler, about 2,500 dollars; derrick and walking beam, 150 dollars; set of drill tools, 300 dollars; hawser, sand pump, drawing pipe, as well as miscellaneous tubes, 1,000 dollars. The contract price for well drilling averages between two dollars and two dollars and fifty cents per foot for the first 500 feet. The total cost of sinking a well 500 feet deep may be estimated at between 6,000 to 9,000 dollars, according to the nature of the ground bored and the strength and careful handling of the machinery. The next following well, subject to parallel conditions, will cost about half as much if the same engine and tools are used. If a flowing well be struck the expenses of operating it are merely nominal. If it be a pump well, two engineers, one or two extra hands, and the fuel, will make the ordinary daily expenses ten to twenty dollars---a variation dependent in a large measure on the price of fuel.

REFINING PETROLEUM.

The process of refining is one which involves a fine combination of the principles of both chemistry and mechanics. The "still" is made of the best iron, and the part of it which is exposed to the fire is made of steel. There are various forms of "stills"; the most generally used resembles a circular boiler. "Stills" vary between 100 and 1,000 barrels capacity, although the latter size is very rare. The still is set in brick-work, arranged to secure a large fire surface and a strong draught. The latest improvement in still-setting presents to the fire half the circumference of the still itself, consisting of a single plate of steel, five-sixteenths of an inch thick.

A quantity of crude oil commensurate with the capacity of the still is put into it and the fire lighted, which is kept burning until the three products of petroleum are separated and drawn off. The still is then allowed to cool, after which a man enters and cleanses it for a new charge of crude oil.

The larger the still-surface exposed to the fire the quicker the products come off. A strong alkali of spent caustic soda is put into the still with the crude oil for the purpose of deodorizing it. The first refined products that come off run for 200 feet or so through piping and a coil immersed in water, for the purpose of cooling them. The first product is vapor, which very soon condenses into a light fluid called naphtha or benzine, and is allowed to run off into a separate tank until its specific gravity reaches sixty-eight degrees, when it is cut off. From that point the product is regarded as refined oil (or kerosene) until the residuum begins to show itself. At this point the "cut-off" is again applied, and the tar (or residuum) is run into a third tank. The refined oil is then pumped into a large tank called the "agitator," where it is thoroughly cleansed from all foreign substances by the steam application of sulphuric acid. After this it is washed with water and run off into a measuring tank, where it is ready for sale and use.

The process of handling refined oil from the time it has left the receiving tank until it is run into the settling or measuring tank is called "treating." The treatment of naphtha is virtually the same as that of refined oil. The residuum requires no chemical treatment except such as it receives while in the still, as a constituent part of the crude oil.

A refinery must be provided with tanks for holding crude oil: with pipes to connect the crude oil tank with the stills; with pumps for drawing the crude oil from tank boats, for forcing it into the stills and for handling the different products while in process of treatment. Also be provided with tanks for receiving the refined oil and the naphtha when they first come from the still, with agitators for cleansing, and with settling or measuring tanks. Wherever feasible, all tanks are placed underground as a measure of precaution against fire, or the spread of it, if fire breaks out above ground.

In regard to the size of refineries as connected with the profits on their products, it may be stated that it costs very little more for the necessary labor to refine 5,000 barrels per week than for that required to produce 500 barrels of refined oil.

The premium on insurance for refineries varies from 25 per cent down to 5 per cent, according to the chances of fire or spreading the same. A wide variety of considerations and a very close scrutiny are employed to determine what these chances really are.

The average rate of wages paid in our country to a competent practical refiner is $17 50 per week in currency. In Europe an equally competent refiner is paid $5 85 in gold. Notwithstanding this great disparity in labor expenses, however, and despite the other fact that great abilities and large capital have been employed to ship crude oil to Europe and to refine it there, the industry of refining does not seem likely to pass away from American control, as many feared it would four years ago.

The total quantity of crude petroleum produced in Pennsylvania between January 1 and December 31, 1873, was 9,867,000 barrels.

In conclusion, it remains only to be noted that the production and stock of crude oil in the oil regions have never hitherto been so large as during the first ten months of this year; that the amount of refined oil in foreign countries was never previously so large as in November, 1873; that the stock of crude oil held by refiners was never before so small as at the date named.

SHIPMENTS OF PETROLEUM FROM NEW YORK.

Appended is a table showing the amount of petroleum and its products shipped from New York to various foreign ports between 1861 (the first year of export) and 1873:

To	1873. Gals.	1872. Gals.	1871. Gals.	1870. Gals.	1869. Gals.	1868. Gals.	1867. Gals.	1866. Gals.	1865. Gals.	1864. Gals.	1863. Gals.	1862. Gals.	1861. Gals.
Liverpool	5,382,539	1,388,419	1,866,558	1,836,675	877,667	1,291,200	1,263,042	2,003,440	1,605,303	734,755	2,156,850	1,781,377	187,254
London	6,233,812	2,763	1,457,628	2,047,118	872,118	947,311	1,599,146	2,835,747	376,283	1,430,710	2,576,381	1,133,399	115,644
Glasgow, &c	105,453				410,605	184,070	192,470	4,189	156,147	368,402	411,943	24,181	276,977
Bristol	1,855,417	556,261	414,520	248,132				155,389	110,412	23,124	71,912		
Hull	74,069	65,814	392,919	83,119									
Falmouth, E., &c	1,228,028	1,021,079		551,649	367,233	98,210	123,933	754,313	509,815	316,402	623,176		
Grangemouth, E.								247,752	102,292		425,384	299,326	
Cork, &c	6,566,273	3,411,436	5,333,811	4,689,233	2,648,865	2,272,534	4,333,150	5,879,863	1,157,486	3,310,362	1,532,257		
Bowling, E.										87,164		195	
Havre, &c	7,994,502	4,139,619	2,832,184	1,417,851	4,275,096	2,925,413	993,274	2,467,482	604,330	2,324,017	1,174,890	794,921	73,716
Marseilles	3,109,258	1,399,830	2,549,793	2,508,468	2,410,308	3,269,600	1,930,181	1,429,753	1,333,752	1,982,075	1,167,893	135,765	1,000
St.Nazaire and Rouen		226,300	309,522			119,450	85,267	100,135	97,841		143,646		
Cette	449,501			108,743						4,800		200	
Dunkirk	1,434,649	850,886	762,369	288,231	831,398	389,401	223,277	63,447	110,699	232,803		2,700	80
Bordeaux and Bayonne	2,168,797	852,292	557,639	455,677	428,306	184,600	104,43	84,929					
Dieppe and Rouen	174,511	229,828		118,772	346,458	78,539	120,453				46,000		
Antwerp	10,518,941	6,489,132	4,747,197	9,977,114	8,202,931	7,052,177	4,721,142	4,220,860	1,739,062	4,119,821	2,692,974	823,090	5,671
Bremen	20,937,777	11,822,831	12,356,572	10,162,299	11,374,282	8,578,026	3,818,671	3,127,562	231,983	971,905	903,004	452,522	32,112
[. . .]	0,904	8,914								77,641	436		
Hamburg, &c	4,727,384	5,776,354	5,866,532	4,456,226	4,333,982	2,458,557	984,689	1,603,484	1,049,300	1,186,680	1,461,155	229,384	42,348
Rotterdam, &c	7,635	1,987,546	6,987,302	5,305,299	2,115,838	1,695,235	2,239,612	537,814	292,569	532,926	757,249	16,938	640
Stockholm and Gott'bg	1,457,412	783,702											
[. . .], &c	6,627,830	3,433,905	5,997,352	7,227,273	4,163,320	1,523,387							
Ana	527,047	397,799	216,047	216,942		150,098							
Konigsberg and Stettin	8,177,785	5,644,478	5,650,978	2,645,677	4,594,363	2,537,086							
Arendal	612,921	143,864	186,260	97,242	138,570			49,730		33,813			
[. . .], &c		294,329										81,960	
Danzig	2,782,669	1,177,776		767,999	810,593	374,671	724,121	944,240	891,359	400,376	88,060		

Copenhagen, Els're, &c...	2,942,388	3,858,708	2,967,345	894,422	341,572	118,492	...	...	97,782	198,538	...	58,474	33,284
Borga, Finland...	339,282	...	73,321	121,540	...	...	...	...	44,988	...	...	16,823	33,000
Sodartolje...	303,426	100,290	91,023	1,848	...	...	...	...	5,128	201,531	...	25,500	...
Syria, &c...	1,261,655	490,520	985,250	287,600	168,220	...	129,253	...	73,751	835,301	...	9,050	308,450	200
Venice...	651,570	...	472,301	06110	...	...	...	...	28,295	89,144	17,474	2,339	...
Cadiz and Malaga...	1,713,013	30,55	84,967	0,049	435,058	380,581	162,251	198,538	22,615	65,780	7,983	57,115	3,990
Tarragon and Alicante...	364,625	35,000	50,760	135,500	66,038	518,260	37,014	...	73,751	835,301	...	308,450	157
Barce'ona...	7,633	774,723	786,685	571,462	530,029	470,929	86,808	201,531	28,295	89,144	9,050	2,339	...
Gibraltar and Malta...	7,599,717	8,023,509	7,397,196	7,932,173	2,774,547	4,289,017	1,466,032	835,301	73,751	1,305,971	17,474	308,450	3,990
Oporto...	241,720	74,590	71,6.0	210,759	362,108	251,704	63,229	89,144	28,295	65,780	7,983	2,339	21,000
Naples and do...	760.5	6,945	870,113	379,912	4,043	1,032,209	210,006	65,780	22,615	915,253	679,666	57,115	62
Genoa and Leghorn...	2,843,463	1,42?,261	3,159,142	2,556,926	1,774,223	2,928	915,253	1,305,971	666,611	49,835	34,075	599,674	...
Trieste...	2,332,953	2,131,130	2,601,290	2,816,655	1,413,743	9,061	246,309	49,835	66,371	13,500	...	3,000	...
Smyrna, &c...	1,3.8,741	1,689,482	1,463,882	1,045,376	748,494	398,873	202,930	13,500	...	...	4,000	...	58
Alexandria, Egypt...	1,515,334	1,932	411,660	8,294	...	223,000	35,776	...	...	...	167,195	64,662	...
Lisbon...	677,397	03,902	07,865	431,582	194,812	43,191	77,091	165,983	93,713	...	3,350	5,125	1,296
...ry Islands...	146,934	61,230	71,865	18,234	16,353	16,461	21,000	10,252	5,244	...	...	...	...
...ntinople...	2,685,030	738,218	1,492,905	1,508,240	602,180	603,012	60,060	4,200	...	...	...	...	...
Bilboa, Seville & Vigo...	1,422,747	1,335,671	2,233,671	2,136,551	1,498,682	417,210	59,939	73,888	...	...	...	...	...
Palm:, Spain, &c...	2,641,728	1,138,408	892,915	935,207	336,221	199,163	...	...	...	...	400	430	...
China and E. Indies...	1,772,239	1,353,030	457,350	451,610	207,150	120,300	108,077	216,551	158,819	...	2,500	...	...
J.pan...	453,850	200,600	169,490	59,272	30,500	24,560	114,540	...	...	...	...	...	...
Afr:a...	477,760	1,990	169,490	1,623,663	619,649	959,959	143,850	96,031	44,630	34,338	36,942	3,970	400
Australia...	2,303,760	1,318,328	47,993	3,680	43,680	37,500	8,000	2,000	17,090	25,195	12,230	655	445
Otago, N. Z...	6,840	133,820	3,680	231,030	139,280	224,520	153,665	56,670	735,891	377,384	394,166	213,699	168,365
Sydney, N. S. W...	8,948	433,644	331,280	1,364,294	835,299	0,096	1,795,542	805,219	14,880	10,310	5,500	7,850	...
Brazil...	1,936,744	2,713,409	1,036,943	243,022	169,541	155,573	96,000	12,600	162,923	97,850	48,013	37,750	5,883
Me...	367,294	382,542	559,809	1,566,547	1,078	988,955	314,015	213,329	291,752	149,676	1,652	54,967	3,712
...ba...	1,734,980	1,850,051	5,751	396,403	101,000	169,200	410,166	528,865	194,936	112,985	9,081	1,816	150,703
Argentine Republic...	603,150	828,573	274,950	7,980	9,020	91,000	4,479	310,645	716,733	418,134	356,436	3856	4,200
...platine Republic...	0,900	529,779	534,050	41,684	193,990	168,000	942,005	1,078,716	68,856	20,260	24,470	7,390	206
Chile...	272,555	270,750	266,360	305,673	21,780	233,956	228,337	184,790	7,252	78,552	71,026	3,327	...
Peru...	12,828	12,462	181,629	5,049	927	1,250	248,100	212,550	53,226	92,550	6,650	17,800	...
British Honduras...	92,365	50,897	8,072	79,543	36,106	8,852	240,706	9070	110,840	169,061	236,007	56,011	...
British ...ua...	...	...	37,150	8,234	...	40,700	3,817	351,090	2,052	6,072	440	...	...
Dutch Guiana...	...	...	11,322	...	...	...	...	5,561	...	...	...	...	...

SHIPMENTS OF PETROLEUM FROM NEW YORK—CONTINUED.

To	1873. Gals.	1872. Gals.	1871. Gals.	1870. Gals.	1869. Gals.	1868. Gals.	1867. Gals.	1866. Gals.	1865. Gs.	1864. Gals.	1863. Gals.	1862. Gals.	1861. Gals.
British West Indies.....	671,192	397,693	489,227	633,492	298,997	236,805	0,500	22,324	800	7,881	15,104	9,396	3,035
Br. N. Am. Colnies....	93,309	69,969	34,930	38,595	54,221	47,521	157,291	3,902	6,941	70,978	60,931	18,888	3,719
Danish West Indies.....	20,330	27,121	10,596	18,058	16,473	12,255	114,029	2,482	4,080	8,902	16,995	2,948	2,636
Dutch East Indies.......	1,330,483	......	277,517	......	......	......	1,468	14,680	10,947	8,463	31,503	4,102	1,770
Dutch West Indies..... .	54,011	48,061	19,823	30,267	40,698	17,463	24,882	22,181	18,369	26,638	12,143	7,117	....
French West Indies	57,240	14,600	3,601	3,600	73,436	77,260	46,225	57,731	52,618	16,020	9,104	2,332	....
Hayti.............	87,421	19,377	40,393	19,634	16,678	3,066	7,238	14,690	13,856	7,088	12,064	1,856	964
Central America.........	82,779	15,465	17,916	3,273	1,858	1,346	5,419	2,566	5,494	993	453	1,764	......
Venezuela.	201,273	132,764	76,020	68,651	77,266	57,911	76,570	58,423	39,794	28,583	15,455	1,094	610
New Grenada.............	104,914	10,478	98,509	78,186	60,312	61,219	83,300	90,718	58,570	57,490	107,887	37,058	15,552
Porto Rico..............	133,240	13,679	93,346	46,934	36,492	34,228	21,899	25,203	42,355	20,026	59,489	20,244	13,925
Sandwich Islands	......	......	......	3,000	......	......	......	......	......	......	......	......	......
Total.........gals..	145,691,935	90,027,726	94,955,850	87,667,299	65,933,690	52,803,202	32,799,120	34,470,061	14,515,173	21,385,784	19,547,604	6,720,973	1,112,476

EXPORTS OF PETROLEUM FROM THE UNITED STATES.

The following table shows the entire exports of petroleum from the United States between 1861 and 1873:

FROM.—	1873. Gals.	1872. Gals.	1871. Gals.	1870. Gals.	1869. Gals.	1868. Gals.	1867. Gals.	1866. Gals.	1865. Gals.	1864. Gals.	1863. Gals.	1862. Gals.	1861. Gals.
New York	145,691,935	90,027,726	94,955,850	87,667,299	65,933,690	52,803,202	33,834,133	34,501,385	14,625,090	21,335,784	...	6,720,273	1,112,476
Boston	2,458,356	1,717,689	2,185,096	1,790,271	2,117,939	2,410,114	2,964,113	1,591,694	1,511,173	1,696,307	2,019,431	660	...
Philadelphia	83,860,120	56,421,900	55,901,590	49,889,736	33,415,552	3,020	29,437,429	28,811,853	12,552,882	7,760,148	5,335,738	2,800,978	...
Baltimore	3,471,222	1,995,104	2,570,538	1,731,321	1,251,423	2,587,707	1,515,454	2,483,419	973,117	929,971	915,866	174,830	...
Pthd.	...	...	...	...	...	705,107	900	12,100	11,088	70,762	342,082	120,520	...
eild.	...	...	...	159,528	...	270,000	...	...	...	...	...	...	...
Total	237,481,633	150,162,419	155,613,064	141,233,155	102,748,604	99,281,750	67,032,029	67,400,441	29,674,350	31,592,972	28,250,721	10,887,501	...
Equal to bbls. of 40 gals.	5,937,041	3,754,060	3,890,326	3,530,068	2,568,715	2,842,041	1,676,301	1,685,011	741,858	783,824	706,268	272,187	...

AVERAGE DAILY PRODUCTION OF PETROLEUM.

The annexed statistics of the average daily production of crude petroleum in the Pennsylvania oil region, for each month since September, 1867, was compiled and published in the Pittsburg Oil Journal:

	1867. Bbls.	1868. Bls.	1869. Bbls.
January		8,700	10,192
February		9,200	9,967
March		8,621	9,891
April		8,537	9,167
May		8,790	0,153
June		0,102	11,334
July		10,593	11,697
August		11,931	12,157
September	9,700	11,033	12,645
October	9,600	0,137	13,071
November	9,800	10,271	3,817
December	10,400	9,730	2,844
Total product for 1868		3,583,176	
Average daily product for 1868		9,811	
Total product for 1869			4,210,720
Average daily product for 1869			11,528

Average Daily Production of Petroleum—Continued.

1870.	Bbls.
January	12,634
February	11,917
March	12,385
April	12,974
May	1,465
June	4,817
July	6,969
August	17,777
Sept mber	19,489
October	20,158
November	18,012
December	15,214
Total product for 1870	5,673,195
Average daily product for 1870	15,543

1871.	Bbls.
January	15,477
February	4,491
March	1857
April	13,308
May	13,987
June	4406
July	17,261
August	18,161
September	17,648
October	,063
November	6,651
December	6,703
Total product for 1871	5,715,900
Average daily product for 1871	15,660

1872.	Bbls
January	16,286
February	17,012
March	5406
April	16,308
May	8,345
June	17,749
July	18,513
August	18,816
September	16,561
October	14,309
November	23,275
December	22,054
Total product for 1872	6,531,675
Average daily product for 1872	17,895

1873.	
January	20,40,
February	21,725
March	21,462
April	21,384
May	25,044
June	26,450
July	27,893
August	3098
September	3019
October	30.403
Total product for first ten months of 1873	7,808,989
Average daily product for the first ten months of 1873	25,677

Year	Average daily product in bbls.	Total product in bbls.
'860	1,320	50,000
1861	5,803	2,118,000
1862	8,373	3,056,000
1863	7,208	2,631,000
1864	5,798	2,116,000
1865	6,841	2,497,000
1866	9,855	3,597,000
1867	9,170	3,347,000
1868	9,818	3,583,110
1869	11,528	4,210,720
1870	15,543	5,673,195
1871	15,660	5,715,900
1872	17,895	6,531,675
1873	25,677	7,808,989

Average daily product of the Pennsylvania oil regions from the discovery of petroleum to November 1, 1873 10,753--

Total product for the same time 53,385,589

COTTONSEED OIL—ANNUAL REPORT

WE are able to present to our readers this year the most complete report ever given of the production of cotton seed oil, we having received returns from all the mills with but one exception. We must thank most of those to whom we applied for information for their prompt replies, and although we have had the figures from all the mills except two or three since before the first of December last, we have delayed our report, hoping to hear from the one still remaining.

We are sorry not to find an increase in the production during the past year 1872-3, the total being 2,304,970 gallons against 2,363,083 gallons the previous season, or a decrease of 58,113 gallons, and an increase of only 245,027 gallons over 1870-1.

Our report of 1871-2 was from 24 mills, which either were or had been running that season, but at the close of the season now under review there were but 13 mills in operation, and only about three others have made any oil since our last report. Many of the smaller works that were not favorably located have been forced to suspend operations, and some of the larger ones have lost money and withdrawn from the business. Two factories have been destroyed by fire.

It may seem quite remarkable that the production the past season has so nearly equaled that of the previous year, when there were nearly twice as many works in operation, but this is explained by the fact that some of the larger mills, particularly in New Orleans, have increased their facilities, and have the capacity for making more oil than they could a year ago. New Orleans and Memphis are still the largest producers.

We do not know of any new uses found for the oil, but it is becoming more favorably known, and consumers are now willing to take cotton seed oil, for many purposes, where they were formerly satisfied that only lard or whale would answer. There is consequently less of it used for adulterating.

Several new methods have been tried for extracting the oil from the seed by chemical means, but as yet, we believe none have been successful. There is said, however, to be a new process found for preserving hulled seed that it may be shipped to all parts of the world without danger of heating. The past season has not been a profitable one owing to the short supply of seed. particularly in those districts where the larger mills were located, making the cost very high on account of the expense of transportation, and prices for other competing oils ruling so low that no advance could be obtained for cotton seed oil.

The following table shows the price of crude, summer and winter yellow for 1872:

		Crude.	Summer Yellow.	Winter Yellow.
January	1	55	61	68
January	15	53	60	65
February	1	51	58	64
February	15	50	56	63
March	1	50	56	62½
March	15	50	57	62½
April	1	50	57½	62½
April	15	52½	57½	63½
May	1	52½	59	65
May	15	52	60	65
June	1	53	59	65
June	15	51	57½	64
July	1	51	57	64
July	15	50	56	63
August	1	50	56	62½
August	15	46	53½	62½
September	1	46	55	64
September	15	46	55	67
October	1	44½	56	66
October	15	47	54	65
November	1	47	53	64
November	15	47½	52	62½
December	1	45	51	62½
December	15	44	51	62½

The following will show the price of oil on the 1st and 15th of each month during 1873:

		Crude.	Summer Yellow.	Winter Yellow.
January	1	44	49	62½
January	15	42½	49	60
February	1	42½	49	60
February	15	44	49	60
March	1	43	48½	58
March	15	43	48	57½
April	1	42½	47½	57
April	15	42½	49	57
May	1	44	49	57
May	15	45	49	57½
June	1	45	51	58
June	15	47½	54	60
July	1	47	54	60
July	15	45	54	60
August	1	43	54	58
August	15	46	56	60
September	1	50	57½	60
September	15	50	58	60
October	1	45	50	60
October	15	45	47½	60
November	1	42	48	58
November	15	39	47½	55
December	1	40	48	54
December	15	40	47½	53

The exports of Cottonseed oil have grown to be of considerable importance, our manufacture taking the preference in Europe, and commanding about £2 per ton more than that made from Egyptian seed. The first exports we note commenced July, 1871, and have continued up to the present time. The ports to which these exports have been made, and the quantity and value to each country are as follows:

1871.

To England	14,953 gallons, valued at	$8,931
Scotland	12,510 " "	7,243
Germany	15,329 " "	8,208
N. G. Union	4,518 " "	2,250
France	189 " "	138
Spain	17,555 " "	8,778
Total for year	65,054	$35,548

1872.

To England	77,319 gallons, valued at	45,190
Scotland	25,404 " '	14,320
Germany	975 " "	487
Netherlands	5,000 " "	2,580
Belgium	456 " "	241
France	664 " "	387
Spain	8,679 " "	4,364
Cuba	12	10
Total for year	118,509 " "	$67,579

1873.

To England	176,241 gallons, valued at	$91,075
Scotland	75,076 " "	39,303
Italy	55,269 " "	29,308
Netherlands	27,130 " "	13,427
Austria	27,156 " "	14,054
British Guiana	225 " "	124
Total for year	361,097 " "	$187,291

The above totals show that the exports in 1872 were 53,455 gallons greater than the previous year, and the totals of 1873 show an increase of 242,588 gallons over 1872.

Statements of receipts and shipments of Cotton Seed and Produce at New Orleans, during the year ending August 31, 1873, compared with the previous years.

RECEIPTS.

		1871-72.	1872-73.
Cotton Seed	bags.	547,678	893,716
Oil Cake	"	82,514	101,549
Oil	bbls.	5,734	5,993
Oil Cake Meal	bags.	729	2,479
" "	hhds.	59	42

		1871-72.	1872-73.
Cotton Seed, New York................................bags.		65,492	40
" Liverpool, (upland)............................ "		22,132	(Sea I) 69.
" Cedar Keys.................................... "		....	1,000
" Vera Cruz .... "		....	1
Total..............................		87,624	1,734

		1871-72.	1872-73.
Oil Cake, Liverpool ....bags.		119,007	*163,319
" London.................................... "		33,390	35,176
" Glasgow .... "		6,571	2,460
·· Leith...................................... ··		4,928	....
" Barcelona................................. ··		51	....
" Charleston, S. C...........................		3,000	....
" New York....................................		4,686	11,006
" Providence, R. I............................. ··		....	647
Total................................		171,633	† 212,608

		1871-72.	1872-73.
Oil—Liverpool.......................................bbls.		2,582	7,157
" London...................................... "		....	350
·· Barcelona.................................. "		6,302	....
·· Cadiz......................................		25	....
" Genoa .... "		20	125
" Havre...................................... ··		205	....
" Havana.................................... ··		2	10
" New York.................................... ··		7,501	14,921
" Providence		5,498	12,061
·· Boston		842	....
" Cincinnati................................... ··		500	....
" Philadelphia................................. ··		100	1,153
Total..............................		23,577	35,780

		1871-72.	1872-73.
Cotton Seed Meal—Baltimore......................bags.		20	....
" Boston .... "		4,148	6,691
" Providence.......................... "		26,655	42,888
Charleston.......................... ··		300	....
Liverpool.		1,011	hhds 75
New York............................ ··		5,925	850
Philadelphia		....	520
Total..............................		38,059	‡ 50,949

		1871-72.	1872-73.
Soap Stock—Liverpool...........................barrels.		35	....
" New York............................ "		790	451
" Havre................................ "		20	....
Total..............................		845	451

---

* 400 bbls.   † 400 bbls.   ‡ 75 hhds.

MENHADEN OIL.

Meeting of Manufacturers--Production, Etc.

FIFTEEN years ago Menhaden, or Porgy oil, as it was then called, was scarcely known in the market. Since that time the manufacture of it has steadily increased, and to-day, in gallons, the production is about equal to the catch of whale oil in this country. And in value the Menhaden interest is ahead of the whale, for though the oil sells at a less price per gallon, for every barrel of oil, over three-quarters of a ton of scrap is made, which sells readily at $15 per ton at the factory, as a fertilizer.

At the first the oil was thick, of a dark color, and generally possessed anything but an agreeable odor. Vast improvements have been made in taking the oil from the fish, and to-day the oil is of a light straw color, perfectly sweet, and can be refined to a state of the greatest purity.

Menhaden oil cannot be used in its natural state for lubricating purposes, owing to its containing so much glutinous matter, but we are assured that a leading Eastern chemist has discovered a process whereby this gum can be separated from the oil, leaving an excellent article for lubricating purposes. What the expense of this treatment is we do not know, but with so many cheap lubricating oils in the market we should hardly think it would pay.

Great improvements have also been made in the methods of taking the fish, as the following letter from a prominent manufacturer describes:

"Since the commencement of the oil business in Narragansett Bay its extension has embraced many localities upon our bays and seacoasts, and many have been the improvements in the method of taking the fish. Originally only the drag seine was used for the purpose, and they could only be taken when near the shore, the great body of the fish keeping in the channel or in deep water. The drag seine is now entirely neglected by the fishermen, and is supplanted by the purse seine. The fishermen now not only resort to the bays and inlets for the purpose of taking their fish, but also to the ocean, where often in favorable weather they take them in large quantities. Our most enterprising fishermen follow them as far East as the coast of Maine, where the fish increase in flesh, and consequently yield a larger quantity of an extra quality of oil. · The fact of the passages having to be made around Cape Cod in open boats, and the fishing season upon the coast of Maine being interrupted with frequent calms and fog, it was extremely difficult for the fishermen to keep their boats together or even to get out to the fishing ground during the continuance of the calm. To overcome such hindrances they have resorted to steam, which has almost done away with the old custom of open boats. Steamboats of from thirty to eighty tons are now generally used by such fishermen as intend to follow the fish along our coasts. The boats and tackle cost when complete from twelve to eighteen thousand dollars each. The steamers are generally owned by the oil manufacturers."

The question has been asked as to how long the fish will continue so abundant, and those who from experience ought to know if there was any likelihood of the supply becoming short say always, that this fish is the bait of the sea, and the young are to be seen in myriads during the fall, and that they will be able to catch fish after ¬etroleum ceases to come from the earth.

At a meeting of the Menhaden Oil and Fish Guano manufacturers of Maine, Long Island, Connecticut, Rhode Island and New Jersey, held in this city January 7th, an association was formed, to be known as " The United States Menhaden Oil and Guano Association," and a Constitution and articles of association were adopted.

The meeting organized with R. L. Fowler, of Guilford, Ct., as Chairman, and Luther Maddocks, of Booth Bay, Me., as Secretary. After some discussion a Committee on Statistics was appointed, with instructions to report as soon as possible. The Committee was as follows: Mr. L. Maddocks, Maine; Mr. Church, Rhode Island; Mr. Price. Long Island; and Mr. Fairchild, Connecticut.

Mr. Fairchild, as Chairman, reported as follows: No. of factories in operation, 62; amount of capital invested, $2,388,000; No. of fishermen employed, 1,197; No. of men employed at factories, 1.109; No. of sailing vessels employed, 383; No. of steamers employed, 20; total number of fish caught, 1,193,100 barrels (250 fish to the barrel); total quantity of oil made, 2,214,800 gallons; total amount of guano made, 36,299 tons; stock in hand of manufacturers, 484,520 gallons oil and 2,700 tons guano.

The meeting then voted to appoint a Committee on permanent organization and to report a Constitution and By-Laws. This Committee consisted of Mr. J. G. Nickerson, Boston; Mr. Thos. F. Price, Greenport, L. I.; and Mr. H. L. Dudley, New Haven. Their report was accepted and the Constitution adopted, and the following officers chosen for the ensuing year:

President. Luther Maddocks, of Booth Bay, Maine; Vice-Presidents, Geo. F. Tuthill, Greenport, Long Island; and R. L. Fowler, Guilford, Conn. ; Secretary and Treasurer, H. L. Dudley, New Haven. Executive Committee: Luther Maddocks, Booth Bay, Maine; David F. Vail, Riverhead, Long Island; B. F. Brightman, Round Pond, Maine.

CONSTITUTION AND BY-LAWS.

NEW YORK, Jan. 7, 1874.

Whereas the manufacture of Menhaden Oil and Fish Guano has become identified as one of the important industries of this country, therefore ·

Resolved, That we, the manufacturers, with the view of rendering to each other mutual aid and assistance, do hereby form ourselves into an association for this purpose, and to be governed by the following constitution :

Article 1.---This association shall be called the "United States Menhaden Oil and Guano Association."

Article 2.—The officers shall be a President, two Vice-Presidents, Secretary and Treasurer and Executive Committee.

Article 3.—The President shall preside at all meetings of the association. In the absence of the President either of the Vice-Presidents may preside. In the absence of all these officers a President shall be chosen *pro tem.*

Article 4.—The Secretary shall keep a faithful record of all business transacted at each meeting of the society, and shall notify members of all meetings by written or printed notice.

Article 5.—The Treasurer shall have charge of all funds belonging to the association, and shall pay them out only by order of the Executive Committee.

Article 6.—The Executive Committee shall consist of three, of which the President shall be one. They shall have power to raise money to meet the expenses of the association by an equitable assessment of each member, and shall have a general supervision of all the affairs and business of the association, not otherwise provided for.

Article 7.—The annual meeting of this association shall be held on the second Wednesday of January, annually. The place of meeting shall be determined by a majority of the Executive Committee, and a notice shall be mailed by the Secretary to each member of the association fifteen days previous to the time of meeting.

Article 8.—Special meetings of the association may be called at any time by the Executive Committee, and upon a written request signed by five members addressed to the President. Notice of all such meetings shall be mailed by the Secretary to each member ten days previous to the time of meeting.

Article 9.—Any person, or any member of any company, engaged in the manufacture of Menhaden Oil and Fish Guano in the United States may become a·member of the association by subscribing to this Constitution and these articles of association.

Article 10.—Each firm or company shall be entitled to but one vote at meetings of the association.

Article 11.—The officers of this association sball be chosen annually by ballot. and shall hold their office for one year or until others are chosen.

Article 12.—This constitution may be amended at any annual meeting, or special meeting called for that purpose, by a two-thirds vote of the members present.

Article 13.—Nine members shall constitute a quoium. but a less number may adjourn.

OILS.

MINERAL OILS FOR IRON.

THE use of heavy mineral oil as a preservative for iron is strongly recommended by the *London Oil Trade Review*, the substance referred to being, we presume, one of the products of shale distillation, so extensively practiced in Great Britain. Whether a similar product can be obtained from our petroleums is a good subject for research. Our contemporary says:

The action of the oil is twofold. First, it is detergent when vigorously and freely brushed over an already rusted surface. It seems to loosen the bulk of the rust and it darkens that which remains. Secondly, it acts as a varnish if applied after the cleansing has been effected, or to new and bright work. Its superiority to vegetable or animal oils depends upon the fact that the bulk of the oil evaporates, and it leaves only a very fine film behind. If the oil is light and fully refined, it evaporates so completely as to do but little good in this way; but if tinged or "once run" oil of sufficiently high gravity be used, the resinous or carbonaceous matter which gives the tinge to the oil remains behind and forms the thin protecting film of varnish. Ordinary varnish leaves far too thick and obvious a film, while the film of the once run oil does its work of protection without displaying itself. As regards the density of the oil required for this purpose, we recommend that which stands between the burning oil and good lubricating oil; it is known, and sometimes sold, as "intermediate oil." We are satisfied that a good trade may be done by anybody who will bring this before the public in a proper manner, and supply the article as required. For domestic purposes, for the cleaning of all kinds of household iron work, for the preservation of such things as mowing machines and other garden tools or exposed iron implements, the brown oil should be sold in small bottles at a cheap rate. For manufacturers of iron-work and for ironmongers, to whom it will prove invaluable, it must of course be supplied in larger parcels. At present, it can hardly be used at all, on account of the difficulty of obtaining it in retail quantities.

LINING BARRELS.

INSTEAD of taking glue which has already been dried and hardened, and then reducing it to a liquid state a second time, E. M. Leggett, of New York, proposes to line oil-barrels with the glue when it is first in the liquid form, before it has been hardened, and thereby save the expense of the usual drying and subsequent liquefaction of the ordinary process.

CONVEYANCE OF OIL IN PIPES.

THE heavy charges for railway freights on crude oils, between the oil regions and the Atlantic seaboard, have led to the preliminary examination of the feasibility of using iron pipes.

The estimated distance from Titusville, Pa., to Philadelphia, allowing 40 miles for undulations of the surface, is 300 miles. The highest elevation required to pump the oil is 1,300 feet. The summit could be reached by means of five pumping stations, each throwing the oil 300 feet high, and driving it eight miles. A six-inch cast-iron pipe is proposed. The capacity for delivery would be over 1,000,000 gallons, or 20,000 barrels per day. The estimated cost is $3,500,000, and the estimated net profits about the same amount, or 100 per cent on the cost.

The piping for crude oil already in operation in the oil districts of Pennsylvania has a total length of 675 miles, inside bore 2 inches, and gathers the oil from the following points to various places of shipment on the Alleghany Valley Railroad :

	Bbls.
Emlenton Pipe Company, from St. Petersburg to Emlenton--capacity per day	1,000
Antwerp Pipe Company, from St. Petersburg to Fullerton--capacity	2,500
Mutual Pipe Company, from St. Petersburg Dist. to Milford, Foxborough, and Clarion —capacity	10,000
Union Pipe Company, from Parker's Dist., etc., to Parker's Landing—capacity	6,000
Fairview Pipe Company, from Argyle, Petrolia, etc., to Brady's Bend—capacity	5,000
Butler Pipe Company, from Grease City to Butler—capacity	1,200

GLASS BARRELS FOR PETROLEUM.

SO much loss has occured from evaporation and leakage in the transportation and storage of petroleum, and its products—benzine and gasoline—that it is proposed to make barrels of glass, sufficiently heavy to stand the pressure and usuage. These, although much more costly at first, would never wear out, and could be cleaned and used for any other fluids indefinitely. An inventor has already obtained a patent for a thick piece of glass inserted into one head of a barrel, and on the part next the bung-hole, to allow the quantity within to be viewed by the buyer.

TESTING OLIVE OIL.

PROFESSOR PALMIERI, of Naples, has lately constructed an electrical apparatus of great delicacy and ingenuity, the object of which is to detect the admixture of other oils with that of the pure olive. The instrument is founded on the fact of the variability in the powers of conduction possessed by the various oils, olive being lowest in the scale. The wires of a battery are brought to a small elongated vessel containing the oil to be examined, and an electrometer being attached, the degree of conductivity can be read off on a scale. The instrument, it is said, can detect any of the usual adulterants with the utmost nicety.

PURIFYING AND SEPARATING FATS.

A METHOD of separating the constituents of animal fat, suggested by Alfred Paraf and used by the Oleomargarine Manufacturing Company of New York, consists in mincing the fat and introducing it, together with its own weight of water, into a wooden tank which is heated by a steam coil to from 100 degs. to 120 degs. F, and constantly stirred. After two hours, the oleomargarine and stearine separate from the scraps and are then allowed to cool, to separate from the water. They are then thoroughly worked with two per cent of salt, put in bags, and subjected to pressure or centrifugal action in a temperature of 60 degs. F, which separates the oleomargarine from the stearine, as the latter is not affected by this heat, although the former is melted by it. After the oleomargarine has again congealed, it may be worked a second time with salt, to separate the last trace of water.

SPONTANEOUS IGNITION OF OILED COTTON OR SILK-WASTE.

MAJOR MAJENDIE has communicated to the Royal Artillery Institution the results of certain experiments, instituted to ascertain the relative degree of risk accompanying the presence of oiled cotton waste and oiled silk-waste in buildings and stores.

M. Galletly, who made the investigation referred to, read a paper at the Brighton meeting of the British Association for the Advancement of Science, on a series of experiments carried on by him, with a view of determining precisely the conditions under which spontaneous combustion takes place in cotton and other combustible material, when impregnated with animal or vegetable fatty oils. Mr. Galletly found that cotton-waste soaked in boiled linseed oil and wrung out, if exposed to a temperature of 170 degs., set up oxidation so rapidly as to cause actual combustion in 105 minutes in the case where the action is slowest. The quantity in this instance was sufficient to fill a box 17 inches long by 17 inches broad, and 7 inches deep, but unfortunately it is by no means necessary that the waste should exist in any such bulk, a common lucifer match-box full igniting in an hour in a chamber at 166 degs. Fahr.

Raw linseed oil ignited less rapidly. The experiment was made in a smaller case than the first one above mentioned. Active combustion took place in four or five hours. Rape oil, and Gallipoli olive oil ignited somewhat less readily, taking at least five hours, though generally a good deal more. Rape oil, in fact, took over six hours at 170 degs. The temperature of 130 degs. was employed in the case of the Gallipoli oil, and also in the following instances: Castor oil took over a day before ignition; lard oil took four hours; salad oil, one hour and forty minutes; and sperm oil refused to char the waste at all.

Mr. Galletly considers that the heavy oils from coal and shale tend remarkably to prevent the oxidation described, by protecting the tissue from contact with the air. It appears that the so-called spontaneous action of oiled cotton waste proceeds from the substance being exposed in a finely divided condition to the oxidizing action of the air. In point of fact, it is the same action that causes the bloom in some of the direct processes for the reduction of iron to revert to the oxide when exposed in a heated state to the air, and the still more remarkable action that is said to have taken place in the iron removed from the Mary Rose. which had lain in the bottom of the sea till it became eaten into a porous condition. It appears to have been hoped that silk-waste might have offered greater security, but this proves not to be the case.

ROSIN OIL SOAP.

ONE hundred pounds of rosin-oil and eighty pounds of lime slaked to a powder are agitated in an iron pot, and the mixture is heated with stirring, till a uniform paste is obtained, free from lumps and running from the stirring implement like syrup. With this rosin oil soap, all the different varieties of patent wagon grease are made as follows:

BLUE PATENT GREASE.—500 lbs. red rosin oil are heated for one hour with two lbs. calcium hydrate, and allowed to cool. The oil is skimmed off the sediment, and 10 to 12 lbs. of rosin oil soap are stirred in till all is of buttery consistence and of blue color.

YELLOW PATENT GREASE is prepared by adding six per cent. of turmeric solution to the blue grease.

BLACK PATENT GREASE.—Lamp-black is used to produce the black color.

PATENT PALM OIL WAGON GREASE.—10 lbs. of rosin oil soap are melted with 10 lbs. of palm oil; 500 lbs. of rosin oil are then added, and as much rosin oil soap to make the whole of the consistence of butter, and lastly 7 to 8 lbs. of caustic soda lye.

PARAFFINE RESIDUES.—The thick oil which remains in the paraffine manufacture is used as a lubricating oil, partly on account of its cheapness, and partly on account of its not soon solidifying by cold.

In order to thicken, some lead-soap is melted with it. Mixtures of rosin oil or rosin oil soap and petroleum, with glycerine also, are often used as lubricants.

ALUMINA SOAP.

BESIDES being applicable to all kinds of material to be rendered waterproof, this soap in solution may be used with advantage to coat metallic surfaces which have to withstand considerable heat. In the air, articles thus coated dry slowly; if exposed to a temperature of 50 degrees. more quickly. To prepare this soap. a solution of alum or aluminium sulphate is added to a dilute boiling solution of soap, so long as a separation of white hydrated alumina soap takes place. The white precipitate is washed with hot water to remove adhering saline solution, and is heated to remove water. By these means a transparent soap is obtained resembling glycerine soap, soluble in warm oil of turpentine in all proportions. The water may also be driven off by heating with oil of turpentine. When the solution becomes thick and transparent, the lake is ready. For making this alumina soap, a good yellow rosin soap is the best to use.

PETROLEUM USED IN MAKING STEEL.

It is stated by J. G. Blunt, of Leavenworth, Kan., that steel may be made from pig or cast-iron, by heating, melting, boiling, and puddling in the usual manner, and after being puddled sufficiently to decarbonize it and remove the impurities, the air-blast is turned off from the furnace, and sufficient petroleum vapor or gas is admitted into the oven to prevent the admission of air. Petroleum vapor is now forced into and through the mass of molten metal in a series of small jets, and by this direct introduction of hydrocarbon the metal can be carbonized to make any grade of steel desired, after which the mass is balled and treated in the usual manner with the squeezers, or hammers and rolls.

PURIFICATION OF LUBRICATING OIL THAT HAS BEEN USED.

We find the following details of a practical method for regenerating lubricating oils given in an Austrian paper: A wooden tub holding 63 quarts has a faucet inserted close to the bottom, and another about four inches farther up the side. In this apparatus is placed 7 quarts of boiling water, in which are then dissolved $4\frac{1}{2}$ oz. chromate of potash, $3\frac{1}{2}$ oz. carbonate of soda, $3\frac{1}{4}$ oz. chloride of calcium, and 9 oz. common salt. When all these are in solution, 45 quarts of the oil to be purified is let in and well stirred for five or ten minutes; after which it is left to rest for a week in a warm place, at the expiration of which time the clear pure oil can be drawn off through the upper stop-cock without disturbing the impurities and cleansing fluid at the bottom.

QUALITATIVE ANALYSIS OF BENZINE.

Commercial benzine often contains quite a large proportion of petroleum, which leaves a disagreeable odor when the benzine is employed for the removal of grease. A small piece of pitch is placed in a test tube and the suspected liquid poured upon it. Pure benzine will readily dissolve the pitch, forming a tarry mass, while adulterated benzine will be less and less colored in proportion to the amount of petroleum contained in it. Coal-tar will dissolve easily in pure benzine, but forms distinct layers when impure material is employed for the solution.

ROSIN OIL AND ITS USES.

ROSIN oil is a product of the dry distillation of rosin. The apparatus used consists of an iron pot, a head piece, a condensing arrangement, and a receiver. In the distillation, a light oil comes over first, together with water. As soon as a cessation in the flow of the distillate occurs, the receiver is changed, and the heat is further raised, when a red-colored and heavy rosin oil comes over. The black residue remaining in the pot is used as a pitch. The light oil, called "pinoline," is rectified, and the acetic-acid water passing over with it is saturated with calcium hydrate, filtered and evaporated to dryness, and the calcium acetate obtained is employed in the manufacture of acetic acid. The rosin oil, obtained after the light oil has passed over, has a dark violet-blue color, and is called "blue rosin oil." The red oil is boiled for a day with water, the evaporated water being returned to the vessel ; next day the water is drawn off, and the remaining rosin oil is saponified with caustic soda lye of 36 deg. Beaume, and the resulting almost solid mass is distilled so long as oil passes over. The product obtained is rectified rosin oil, which is allowed to stand in iron vessels, protected by a thin layer of gypsum, whereby, after a few weeks, a perfectly clear oil is obtained free from water. The oil of first quality is obtained by a repetition of the foregoing operation upon the once rectified oil. The residues of both operations are melted up with the pitch.

VULCAN.

" VULCAN OIL " is the distillate of Virginian petroleum which passes over almost at the last, and has a specific gravity from 0.870 to 0.890. This, while warm, is acidified with 6 per cent. of fuming sulphuric acid in a lead vat, then drawn off from the acid, and washed with water to the complete removal of the acid. The product is then mixed with 5 per cent. of rape oil. Also that oil which distills over in the petroleum distillation after the illuminating oil (until of specific gravity 0.860) passes over, is taken separately, until it solidifies when dropped upon a cold metal plate, and with this 5 per cent. of crude rape oil is mixed.

Another lubricating oil from petroleum is the " opal oil." This oil, of specific gravity 0.850 to 0.870, is acidified like "vulcan oil," and mixed with 10 per cent. of rape oil.

DEODORIZING OIL FROM ACID TAR.

RICHARD GAGGIN, of Erie, Pa., states that a good oil suitable for lubricating purposes, and for use as a substitute for linseed oil in the manufacture of printers' ink, may be obtained from the acid tar of oil-refineries, by diluting it with benzine, then separating the acid by repeated washings, distilling, and next treating with milk of chloride of lime at a temperature not exceeding 140 deg. After the oil has been thus treated, the limy sediment is drawn off, and a caustic or carbonated alkali intro-duced to neutralize any of the remaining chlorides or chlorine. The alkaline sediment is next drawn off, and after the oil has been again washed with water, the process is finished.

CLEANING GUNS WITH PETROLEUM.

GREASING a weapon with fats and oils does not entirely protect it from rust; the so-called drying oils get gummy and resinous, while the non-drying oils become rancid; and by exposure to the action of the atmos-phere, acids are formed, and these attack the iron. These are some of the reasons why petroleum is to be preferred for this purpose. Petroleum is as great an enemy to water as are the fatty oils; and hence when a gun-barrel is covered with a film of petroleum, it keeps the water away from the metal which forms the barrel; the water which rests upon this film of petroleum evaporates, but the oil does not, and hence no rust can be formed. It is very essential, however, that the petroleum or kerosene employed be perfectly pure, for impure oil, such as is often met with in commerce, attacks the metal. Care must also be taken not to allow it to come in contact with the polished stock.

The gun is cleaned as follows: Each rifleman carries a tin flask of pure kerosene and a round brush, of stiff hog's bristles, which fits the barrel of the gun. The brush is screwed to the ramrod. The gunner also carries some dry hemp or tow. When about to clean a gun, some tow is wrapped about the rod and enough petroleum poured upon it to thoroughly moisten it; it is then pushed in a rotary manner through the barrel and back a dozen times, and the hemp taken out and unrolled, and the upper and lower ends of the barrel rubbed with the clean part, after which it is thrown away. This removes the coarser portion of the dirt. The brush is then moistened thoroughly with petroleum and twisted into the barrel, running it back and forth at least a dozen times, thus loosening the dirt that is more firmly attached to it. The first operation is now repeated, except that the hemp or tow on the rod is left dry, and the rubbing with this must be continued in all directions as long as it comes out soiled. The use of wire brushes is objectionable for cleaning guns, as the numerous little steel points cut into the tube. Only soft tow, hemp, woolen rags, or the like, should be used, as the petroleum dissolves off the dirt sufficiently.

BLACK OIL.

THIS name is applied to a mineral oil produced chiefly in West Virginia. It is also called West Virginia oil, rock oil and lubricating petroleum. There is no branch of the oil trade in which there are larger single corporations, or greater capital employed than in West Virginia oils. The introduction of these oils on the market is comparatively recent, and the enormous amount, now produced and sold, proves it to be a good lubricator.

It is said that many parties formerly using only winter bleached sperm oil now use this oil and consider it quite equal in all respects. A great deal that is interesting might be written with regard to this oil. It may be well, however, to add that there are adulterations sold as pure West Virginia, composed chiefly of petroleum residuum.

A well-informed correspondent in West Virginia sends us the following. The probable daily production of the Volcano region is about 150 barrels of 30 deg. and heavier; about 225 barrels of $30\frac{1}{4}$ to $31\frac{1}{4}$ deg.; about 400 to 500 barrels from $30\frac{1}{4}$ to 35 deg. Burning Springs produces about 150 barrels per day of 37 to 40 deg. The Cow Run (Ohio region) about 200 bbls of 40 deg. and lighter gravity. All this territory embraces about thirty-five miles, but the regions are only from three to four miles square.

OIL DEPOSITS OF THE WEST.

About eight hundred miles west of Omaha, the line of the Union Pacific Railroad crosses Green River, and the approach to the river is for a considerable distance through a cutting of from 20 to 40 feet in depth, made in rock. During the construction of the road, some workmen piled together a few pieces of the excavated rock as a protection for a dinner fire, and soon observed that the stone itself ignited. The place thereafter became known as Burning Stone Cut.

The general superintendent of the road, Mr. T. E. Sickles, has caused analyses and experiments to be made with this substance, which proves to be a shale rock, rich in mineral oils, which may be produced by distillation in abundant quantities, say thirty-five gallons to the ton of rock, at the cost of a few cents only per gallon. The oil thus obtained is of excellent quality, and comes over in two or more grades, one suitable for burning and one for lubrication. Its abundance and cheapness of production is such as to render it certain that the markets of the Pacific coast, and all places west of the Mississippi will, ere long, be wholly supplied from these deposits. The oil can be distilled, delivered and sold at the points indicated at cheaper rates than the Pennsylvania and West Virginia oils can be transported to the Mississippi.

The deposits in question are supposed to cover an area of territory one hundred and fifty miles long and fifty miles broad. They overlie the immense coal-beds found in that region, and consist of sandstone impregnated with oil. They are supposed to have originated by the absorption of oil by sand, the oil having been expelled from the ancient vegetable growths by heat and pressure during the original process of coal formation.

W.G. Jackman.

FISHER HOWE.

NEW YORK CITY and Brooklyn possess some of the noblest religious and charitable institutions in the world. They have been organized, for the most part, within the last fifty years, and are the monuments of a class of men these cities have produced. In connection with the most intense devotion to business, many of the merchants of these two cities have given their time and wealth to the establishment of Hospitals, Asylums, Aid Societies, Seminaries and organizations for relief of all kinds. The title " Christian Merchant " has been given to them, because of their zeal in these two departments of commerce and philanthropy. Without entering into politics or even gaining a wide personal reputation, they have yet been instrumental in making our two cities an honor and a pride to us all. For the institutions of religion and education and philanthropy are, after all, the chief glory of any city or nation.

MR. FISHER HOWE was a merchant of this class. He was emphatically a christian man, and his life was devoted to efforts for the benefit of others. He was, also, like most men of this class, *self-made*, for he came to New York in 1820 with but $50, and afterwards maintained himself and earned his property, by his own endeavors.

. He was born in the town of Rochester, New Hampshire, September 3, 1798. His father, the physician of the place, died when he was nine years of age, leaving his family, with their widowed mother, to the care of his oldest son, James. To this brother the young boy Fisher was greatly indebted, and in after life, he made record of his gratitude, in these words: "To him, under God, I owe all that I am and have." Soon after the father's death, the family moved to Haverhill, Mass., where the lad Fisher was to begin his knowledge of mercantile life. He there entered the country store of his brother, James How. In this store, as was the custom of the time, he had to perform the menial services, and afterwards would often speak of being up early and late, and sweeping out the store, and making fires in those early days. He began, while in Haverhill, to read books of a profound and serious character; a style of reading which always interested him. His mind was particularly eager in its search after the knowledge of books, and it is evident from the taste he cultivated by his own efforts, that he would have made no ordinary scholar if he had been permitted a college education. But he was not able to go to college, and hardly able to obtain a district school education.

In June, 1820, he started from Boston for New York, and on his arrival, rented a part of the store, 171 Pearl Street, at $375 a year, and engaged board and lodging at $3.50 per week. In August of the same year, he formed a partnership with his brother, Calvin W. How, and J. G. Ward, under the firm name of Ward, How & Co., for the transaction of the boot and shoe business. Mr. Ward soon left the firm, and Mr. David Wesson became associated with the brothers How, the new firm bearing the name C. W. How & Co.

It is interesting to discover the following desire entered in his diary at the commencement of this business life: " May I find that in all my connections and conduct, integrity and uprightness may preserve me." This desire seems to have been his motto, for he was always remarkable for strict integrity and uprightness.

Soon after coming to this city, Mr. How joined the Brick Church, then under the pastoral care of the Rev. Dr. Spring.

He had not been long a member of this church before he was elected an elder. This position he filled at the early age of 28 years. It is an evidence of the respect the stranger youth had won in the city that so influential a church as this should choose him to one of its most important offices. The health of his wife*, to whom he had been married since his coming to New York, obliged him in 1830 to leave his residence, No. 81 White Street, and move to Brooklyn Heights. He there began the erection of a house in Willow street, which he occupied for forty years to a day,—moving in November 8, 1831, and dying there November 7, 1871. When he built the house, it commanded an unobstructed view of New York, Staten Island and the Narrows. It was built with the improvements then modern, and had the first arrangements of water, tank and pump, and also the first speaking tube used in the city.

His wife died very soon after her removal to Brooklyn.

In 1832 (Oct. 1) Mr. How was married to the daughter of David Leavitt, the well-known banker, who still lives hale and hearty, although more than four score years of age. At the time of his marriage he added the letter *E* to his name, and afterward wrote his family name HOWE, instead of HOW. This he did in the belief that HOWE is more in accordance with the early usage of the family than HOW.

The Brooklyn White Lead Co. was organized under the Presidency of Mr. Leavitt, in 1825. MR. HOWE'S connection with the Company dates from 1842. He then became its Treasurer and General Manager, a position which he retained, with a single brief interval, until the time of his death. As Manager of the Company, he gave shape and prosperity to the business, as though it had been his private property. He was always industrious, pains-taking and considerate of his associates and employés. He invented and put into operation a plan

---

* Mr. H. was married to Matilda Saltonstal, daughter of Dr. Saltonstal, of Haverhill, Mass., June 16, 1825.

by which the workmen in the lead works are protected against
the poisonous dust, so fatal to health and life. He also did
very much towards founding the organization, known as "The
Association of American Manufacturers of White Lead." He
was the President of the Association until the time of his
death. The following minute was adopted by the Association
when his death was announced:

"We desire to record our high appreciation of his many virtues and
the esteem in which he was held by us.

"As our presiding officer, he was dignified and courteous; as our
counsellor, wise and sagacious; as our associate, a Christian gentleman and
friend.

"For his memory, we shall ever entertain feelings of the highest regard
and affection."

"The Western Lead Corroders' Association" passed the following minute:

"It is with sincere regret that we have heard of the death of FISHER
HOWE, Esq., President of 'The American Lead Corroders' Association,'
a successful manufacturer of marked ability, an affable and courteous
gentleman. We revere his memory, and shall strive to emulate his
virtues. We desire, in this manner to express to his associates in business
our sympathy for the loss of so able an adviser and friend."

It is believed that the organization he formed has proved of
great benefit to the White Lead business, by arranging the
prices of lead, and creating a sympathy and acquaintance
among members of the trade.

When Mr. HOWE left the lead business soon after he first
became treasurer of the Brooklyn White Lead Company, he
formed a partnership with Abijah Fisher and J. C. Hamilton,
under the firm name of Fisher, Howe & Hamilton, for the
transaction of the jobbing and general commission business.
They were agents for the famous New York Mills. This partnership did not continue long, however, for soon he was wanted
in his old place again. He went back as treasurer and manager,
and remained in the position as long as he lived. Any one
at all familiar with the lower part of Maiden Lane, these last

ten years, will recall his venerable appearance and feeble step as he entered his office at about ten in the morning and left it by two in the afternoon. He was reluctant to give up his active life, and performed his duties long after his strength and health had failed. In 1849 Mr. HOWE made a tour of Europe in company with his wife and niece, and in the spring of 1850 they extended their journey to Syria and Palestine. He was a careful and observant traveller. His guide books were always open, and very little escaped his notice. Twenty years after his return he wrote to a friend to visit certain localities in and about Jerusalem, and confirm his impressions of the formation of the hills. And the friend found that these impressions, twenty years old, corresponded with the general contour of those sacred hills. Mr. HOWE was an enthusiast in Oriental exploration. He would lay aside almost any work to discuss some question of topography; and as he talked his eyes would brighten and his speech quicken and glow until he became fairly eloquent. When he came back in 1850 he prepared a volume and published it, entitled: "Oriental and Sacred Scenes," and just before his death he received from the press a smaller work he had written on "The True Site of Calvary."

Engaged in his studies and in acts of religious duty and benevolence, Mr. HOWE's life passed very pleasantly and profitably. He acquired some knowledge of the Hebrew, Greek and Latin languages, and devoted much time to the study of the Sacred Scriptures. He was emphatically an admirer of the Bible. He read it constantly and never questioned its teachings nor hesitated to go where those teachings seemed to point him.

His home in Brooklyn became a familiar resort to Christian missionaries, ministers and philanthropists. Many foreign missionaries have sailed to their stations from his house, and they can remember how carefully he provided for the comfort of their voyages. He gave regularly and conscientiously to the various benevolent societies. When he went to Brooklyn he connected himself with the First Presbyterian church in Henry street, and was soon elected an elder there. He was called by his associates, in later years, "the beloved disciple," on account of the amiability and gentleness of his disposition.

As his infirmities increased, he was less frequently seen abroad; yet many of his fellow-citizens and friends sought his counsels at his home, and enjoyed his society there.

On the 7th of November, 1871, at the ripe age of 73 years, he died. He was in full possession of his faculties to the last. He was in his own house. His wife and his children, with a single exception, stood about him. He was at peace with mankind and with God. His last words to those near him are perhaps as characteristic of his life as any could be :

"Let my children live in peace, and the God of peace will bless them."

It is, of course, impossible to estimate the results of a life like this of Mr. Howe. Passed in retirement, its actions are not noised abroad; yet by the multiplication of such lives and efforts our social, educational, and political, affairs are wisely regulated.

As a merchant, Mr. Howe was remarkable for his integrity. He was emphatically an honest man. He labored industriously, yet never to the neglect of home pleasures and the cultivation of his mind. During all the years of his business life he was accustomed to dine with his family at three o'clock, and after that to remain with them, either driving in the country about Brooklyn, or else enjoying the quiet of his own house. He did not covet public positions of mercantile trust. He served with fidelity and acceptance as a director in the American Exchange National Bank from June 3d, 1837, and was connected with the management of the Brooklyn City Railroad from its commencement. But he preferred his own business and his privacy to such official work.

He was a cautious man, and even timid, in carrying out his plans, which were sagacious and thoughtful. Perhaps this was owing to his feeble health, for he was always frail and delicate;

and in the year 1842, when he spent the winter in Santa Cruz, no one thought that he would live long.

As a citizen of Brooklyn, he was honored and beloved. He represented the first ward for one or two terms as alderman; he was a member of the Board of Education, and was always punctual and regular at the meetings of the Board and in the school visitation; he lent his aid also to the founding of the city hospital. Says the Hon. C. P. Smith, of Brooklyn:

"In the early organization of Greenwood cemetery the Trustees had contracted for a considerable tract of land, the most important for the cemetery. After a few years the Trustees found themselves greatly embarrassed to meet the payments of the land, and stopped selling burial lots, and solicited the aid of the churches. Christ Church and the First Presbyterian Church came to their aid. Christ Church responded by the purchase of about $2,000 worth of lots, and the First Presbyterian Church gave $6,000 by issuing their bond with interest, agreeing to confine the sale of lots to members of the congregation. The owners of the tract of land took the bond of the church and conveyed the land to the cemetery. In this movement Mr. HOWE was active and very influential in bringing this about. He with me agreed to indemnify the church from loss. The Greenwood cemetery and the church have reason to be grateful to the memory of Mr. HOWE."

Mr. Smith also adds, in speaking of Mr. HOWE as a citizen:

"He was intelligent, active and ever entirely reliable. He was connected with and active in all charitable and religious organizations intended for the relief of the ignorant and poor."

As a man, Mr. HOWE was a Christian gentleman. He had the mind of a student, and the cultivation which books and travel and good society give. In his home he gathered pictures and flowers and curiosities from every land. He was quiet, pure-minded, and possessed of a singularly devout nature. The religious element seemed to predominate with him. As a consequence, he was best known in connection with the church and the works of the church. He was one of the founders of the Union Theological Seminary in New York, and took a lively interest in its welfare. He joined the Oriental Society, because he felt that it would assist the understanding of the Bible, and

for the same reason was an active member of our American Palestine Exploration Committee.

The two books he published had the same design. His first
volume, " Oriental and Sacred Scenes," is thoughtful, discrimin-
ating and eminently reverent. Those who have made it a hand-
book of Palestine travel know that it is most suggestive and
accurate. His second and last book, " The True Site of Calvary," discusses the vexed question of the Holy Sepulchre of
Jerusalem. Denying the correctness of the present site within
the city walls, Mr. Howe asserts that we may look for the true
site on the skull-shaped mounds over the cave of Jeremiah.
This assertion he supports by arguments drawn from Scripture,
from history, and from the topography of the country. Many of
the first scholars of our country expressed their interest in this
investigation, and Mr. Howe's treatment of the subject, and not
a few of them acknowledged that they were convinced of the
correctness of his theory.

It is not often that an ordinary merchant, without the advantages of a college education, is able to lay before the public
two volumes so valuable to the cause of truth.

On the day of his death the Brooklyn *Union* contained this
tribute to his worth :

" Thousands of our oldest citizens will be deeply pained to hear of the
sudden death, this morning, of FISHER HOWE, Esq., at his residence in
Willow street. in this city. Mr. HOWE has been a resident of Brooklyn
for a period of nearly forty years ; and has been, until recently, when
health failed, one of the most active and useful men among us. He was
consistent, liberal, devoted, and exemplar, as a church member ; honest,
faithful, and patriotic as a citizen ; kind, affectionate, and devoted as a
husband, father, and friend ; and, on the whole, was one of the most useful
men in the city. He was always connected with our best and most prominent institutions, and was ever prompt and faithful in the discharge of his
public duties. Mr. HOWE was also a clear-headed, successful business
man, and was a wise counsellor and a judicious manager in all financial
matters. He was a man of culture, and of considerable literary attainments. He always took a deep interest in the dissemination of gospel
truth in distant parts of the world ; was a devoted friend and liberal sup-

porter of missions. He knew the field of missionary labor perhaps as well as any man living, disconnected with missionary institutions.

"He was extravagantly fond of the beautiful in nature—of beautiful fields, flowers, and of everything beautiful in art, pictures, statuary and architecture.

"But his character as a Christian was most noticeable and most admired. He was firm, yet cheerful; modest, yet zealous. He was devoted to his church, to his pastor, to the whole congregation with which he so long worshipped, and he will be missed in death probably more than any other member connected with the First Presbyterian Church—Dr. Seaver —in Henry street. Good men, like FISHER HOWE, are scarcer than gold in any community, and are worth more than fine gold and precious stones, wherever they are permitted to live and labor. His was a living example for good, known and read of all men with whom he came in contact, and although dead, his noble Christian example will long speak to his praise, speak to encourage the host of friends he left behind, and speak only for their best good, present and future."

Mr. HOWE left a widow and six children and many grand-children to revere his memory and mourn his loss.

WHITE LEAD.

THERE is probably no country in the world where more of this paint is used, in proportion to the population, than in the United States. Most of the white lead used in this country is of home production, owing to a protective tariff of three cents per pound. We import a considerable quantity from England, and a little from Germany. The foreign lead is whiter, though the American has more body. The corroders of New York, Boston and Philadelphia use almost entirely imported pig lead, but the Western manufacturers use the American. The following is an estimate of the production by one of the most competent manufacturers in the trade:

For 1873 I think the quantity of white lead made was about 29.000 tons east of the Alleghenies; 16,500 tons west of the Alleghenies; making 45,500 tons of 2,000 pounds, equal to 91,000,000 pounds of dry white lead. To grind pure this would require about 10 pounds of oil to 100 pounds of lead, or 1,213,333 gallons of linseed oil, viz., 7½ pounds each—one million, two hundred and thirteen thousand, three hundred and thirty-three gallons oil.

Dry lead as above,	91,000,000 pounds.
1,213,333 gallons of oil as above,	9,100.000 "
Strictly pure white lead in oil,	100,100,000 "

Had the different factories been pushed the whole year, a larger quantity could have been made—probably 10 to 15 per cent more.

The process of making the white lead, though tedious, is not by any means a difficult and complicated one, and, what is perhaps singular, it has varied very little in many years. It consists essentially in exposing thin sheets of lead in gallipots to the vapors of diluted acetic acid, the gallipots being at the same time closely covered and surrounded with spent tan or some similar substance. In a few weeks the lead will be found to have been entirely converted into the carbonate (white lead of commerce), though the precise chemical reactions which take place are not very clearly understood. Another process has been proposed and used to some extent, though without satisfactory results, whereby the carbonate was obtained by subjecting litharge to the action of acetic acid, and then to a stream of carbonic acid gas. The old process is the one in use by our largest manufacturers.

Either of the processes described furnishes dry lead, which has then to be ground with linseed oil before it can be used as paint. It is in this grinding that the lead is adulterated, the chief articles used for that purpose being the sulphate of baryta and zinc. These adulterations are, of course, well understood, and there does not seem to be any essential lack of honesty in them, though the larger companies prefer to make only pure lead, and to sell none other in oil. The dry lead is, however, bought by houses engaged in grinding only, and by them sold in brands of different purity and corresponding prices.

It is stated by the dealers that while there is a large demand for the lower grades for frontier towns and newly settled portions of the country generally, the progress of a town in wealth and importance is invariably followed by a demand for a better quality of paint, so that in our large eastern cities the retail sales of adulterated lead are comparatively nominal. The reason is obvious. For cheap work and temporary structures the adulterated paint is quite good enough; while more permanence of structure demands a corresponding improvement in all the articles used in building.

PAINTS.

PREPARATION OF LIGHT DRYING VARNISH.

TWENTY-FIVE pounds of pure linseed oil are poured into an enameled iron pot, which holds about forty pounds weight; the pot is then placed on a moderately strong charcoal fire, and the linseed oil heated for about half an hour to the boiling point. In the mean time four ounces of pure oxide of manganese are to be rubbed down in linseed oil. This mass is then put into a small vessel provided with a spout, and poured in drops into the boiling linseed oil, while being gently stirred with a wooden spatula.

During the rising and effervescence of the heated oil, the dropping in of the manganese preparation must stop.

As soon as the oil has settled, the dropping in is continued to the last. The vessel is washed out with linseed oil, which is poured into the boiling oil. The varnish is now boiled slowly for an hour, but if a stronger or more quickly drying varnish is desired, it should be boiled for half an hour or an hour longer.

The finished varnish is then removed from the fire, covered with a clean plate, and left to rest for about twenty-four hours, then carefully poured off into clean vessels. The sediment and other residue are generally used for ordinary ground colors.

The pure linseed oil varnish poured into glass bottles can be perfectly bleached by selecting a suitable spot where the sunlight and moonlight penetrate. According to Dr. Gromann, moonlight bleaches quicker than sunlight. The clear bleached linseed oil varnish is used only for the finest white oil and lac colors, and for dissolving the copal lacs, as well as a drying medium for all fine oil colors.

ZOPISSA.

THIS is essentially a paint composed of boiled linseed oil, brown umber, lime water, sulphate of copper, Prussian blue, copperas, burnt clay, calcareous silex, (whatever that may be,) litharge, asphalt, red-lead, gum animi, and turpentine. It was probably through mere modesty that the inventor stopped after adding these ingredients and did not continue through the drug shop. The paint is no better and no worse than one containing an impure oxide of copper for pigment, and linseed oil and asphaltum for the menstruum. It will no doubt protect the wood to which it is applied in sufficient quantities, from external action, so long as it lasts.

BLACK VARNISH FOR ZINC.

PROFESSOR BOTTGE prepares a black coating for zinc by dissolving two parts of nitrate of copper and three parts crystallized chloride of copper in 64 parts of water, and adding eight parts of nitric acid of specific gravity. This, however, is quite expensive; and in some places the copper salts are difficult to obtain. On this account, Puscher prepares black paint or varnish with the following simple ingredients: Equal parts of chlorate of pota h and blue vitrol are dissolved in 36 times as much warm water, and the solution left to cool. If the sulphate of copper used contains iron, it is precipitated as a hydrated oxide and can be removed by decantation or filtration. The zinc castings are then immersed for a few seconds in the solution until quite black, rinsed off with water, and dried. Even before it is dry, the black coating adheres to the object so that it may be wiped dry with a cloth. A more economical method, since a much smaller quantity of the salt solution is required, is to apply it repeatedly with a sponge. If copper colored spots appear during the operation, the solution is applied to them a second time, and after a while they turn black. As soon as the object becomes equally black all over, it is washed with water and dried. On rubbing, the coating acquires a glittering appearance like indigo, which disappears on applying a few drops of linseed oil varnish or " wax milk," and the zinc then has a deep black color and gloss. The wax milk just mentioned is prepared by boiling one part of yellow soap and three parts Japanese wax in 21 parts of water, until the soap dissolves. When cold, it has the consistency of salve, and will keep in closed vessels as long as desired. It can be used for polishing carved wood-work and for waxing ballroom floors, as it is cheaper than the solution of wax in turpentine, and does not stick or smell so disagreeable as the latter. A permanent black ink for zinc labels is prepared by dissolving equal parts of chlorate of potash and sulphate of copper in 18 parts of water, and adding some gum arabic solution. The black polish above described is recommended as permanent and capable of resisting quite a high temperature.

WHITE PAINT FOR METALLIC SURFACES.

DR. SELS says when oil paints are used for metallic surfaces that are subjected to heat, they turn yellow and brown from the burning of the organic portion of the paint. If, instead of oil, soluble glass be used, there will be no organic or combustible substance to brown it. Zinc-white mixed with soluble glass of from 40 deg. to 50 degs. B., to the consistency of ordinary paint, forms a beautiful and permanent color that will stand any required heat, without browning and blistering, and can only be removed by mechanical means. A not very large quantity should be mixed at one time, as a chemical change takes place and the paint hardens.

PAINT ON WROUGHT IRON PLATES.

THIS matter has been carefully investigated under the auspices of the Dutch State Railways, and the result was brought before the Society of Dutch Engineers by Van Diesen. Of thirty-two plates, half the number were plunged in diluted hydrochloric acid for twenty-four hours, then neutralized with lime (slaked), rinsed in hot water, and while warm, rubbed with oil; the other half were mechanically cleaned by means of scraping and brushing. Four plates of each kind were then prepared with one coat of red lead, two sorts of oxide of iron, and coal tar. This was done at Harkort's works in July, 1867. The plates were then exposed to the influence of the weather for a period of three years and re-examined, when it was found: 1. That the coating of red lead had stood well on plates of either method of preparation, therefore in this case no preference could be assigned. 2. That oxide of iron, by Kampand Soeten, gave better results on plates cleaned chemically than on those cleaned mechanically, the coat on the former being in as good a state of preservation as the red lead. 3. That oxide of iron by Anderghem gives as good results on plates cleaned chemically as the two previous materials, but is inferior to them if applied to scrubbed surfaces. 4. That coal tar is much inferior to any of the preceding; from scrubbed surfaces it had almost entirely disappeared.

WATERPROOF PAINT FOR CANVAS.

THE following is a cheap and simple process for coating canvas for wagon-tops, tents, awnings, etc. It renders it impermeable to moisture, without making it stiff and liable to break. Soft-soap is to be dissolved in hot water, and a solution of sulphate of iron added. The sulphuric acid combines with the potash of the soap, and the oxide of iron is precipitated with the fatty acid as insoluble iron soap. This is washed and dried, and mixed with linseed oil. The addition of dissolved india-rubber to the oil improves the paint.

PREPARING WATER-COLORS.

A CONVENIENT method of preparing water-colors for artistic and other uses, suggested by E. L. Molyneux, consists in preparing a sheet of cardboard in such a manner that it will not absorb color, and then painting it with any desired color prepared with sufficient sizing. When one coat is dry, another is added, and so on, until a mass of sufficient thickness has been formed. These sheets are cut up and the pieces may then be pasted on different sheets, each sheet containing as many different colors or pieces as desired. In using these tablets, the amount of color required may be taken off with a brush moistened with water, thus dispensing with the use of the slabs or tiles usually required in grinding the cakes.

LACKERS.

LACKERS are used upon polished metals and wood, to impart the appearance of gold. They are wanted of different depths and shades of colors. The following are recipes for their preparation:

1. DEEP GOLDEN-COLORED LACKER.—Seed-lac, 3 ounces; turmeric, 1 ounce; dragon's blood, ¼ ounce; alcohol, 1 pint. Digest for a week, frequently shaking. Decant and filter.

2. GOLDEN-COLORED LACKER.—Ground turmeric, 1 pound; gamboge, 1½ ounces; gum-sandarac, 3½ pounds; shellac, ¾ pound (all in powder); rectified spirits of wine, 2 gallons. Dissolve, strain, and add one pint of turpentine varnish.

3. RED-COLORED LACKER.—Spanish annatto, 3 pounds; dragon's blood, 1 pound; gum-sandarac, 3¼ pounds; rectified spirits, 2 gallons; turpentine varnish, 1 quart. Dissolve and mix as the last.

4. PALE BRASS-COLORED LACKER.—Gamboge, cut small, 1 ounce; Cape aloes, ditto, 3 ounces; pale shellac, 1 pound; rectified spirits, 2 gallons. Dissolve and mix as No. 2.

5. Seed-lac, dragon's blood, annatto and gamboge, of each ¼ pound; saffron, 1 ounce, rectified spirits of wine, 10 pints. Dissolve and mix as No. 2.

The following receipes make most excellent lackers:

1. GOLD LACKER.—Put into a clean four-gallon tin 1 pound ground turmeric, 1½ ounces of powdered gamboge, 3½ ounces of powdered gum-sandarac, ¾ pound of shellac, and 2 gallons spirits of wine. After being agitated, dissolved, and strained, add 1 pint of turpentine varnish well mixed.

2. RED LACKER.—Two gallons spirits of wine, 1 pound of dragon's blood, 3 pounds of Spanish annatto, 4½ pounds of gum-sandarac, 2 pints of turpentine. Made as No. 1 lacker.

3. PALE BRASS LACKER.—Two gallons spirits of wine, 3 ounces of Cape aloes cut small, 1 pound of fine pale shellac, 1 ounce of gamboge, cut small, no turpentine varnish. Made exactly as before.

But observe that those who make lackers frequently want some paler, and some darker, and sometimes inclining more to the particular tint of certain of the component ingredients. Therefore, if a four ounce vial of a strong solution of each ingredient be prepared, a lacker of any tint can be procured any time.

4. PALE TIN LACKER.—Strongest alcohol, 4 ounces; powdered turmeric, 2 drachms; hay saffron, 1 scruple; dragon's blood, in powder, 2 scruples; red saunders, ½ scruple. Infuse this mixture in the cold for forty-eight hours, pour off the clear, and strain the rest; then add powdered shellac, ½ ounce; sandarac, 1 drachm; mastic, 1 drachm; Canada balsam, 1 drachm. Dissolve this in the cold by frequent agitation, laying

the bottle on its side, to present a greater surface to the alcohol. When dissolved, add 40 drops of spirits turpentine.

5 ANOTHER DEEP GOLD LACKER.—Strongest alcohol, 4 ounces; Spanish annatto. 8 grains; powdered turmeric, 2 drachms; red saunders, 12 grains. Infuse and add shellac, etc., as to the pale tin lacker; and. when dissolved, add 30 drops of spirits of turpentine.

Lacker should always stand till it is quite fine before it is used.

WOOD-STAINING.

1. ORDINARY BLACK STAINING.—Brazil wood, powdered gall-nuts, and alum are boiled in water until a black color is obtained; the liquid is filtered and spread upon the wood, which is then covered with a prepara- tion of iron filings heated for some days with vitriol and vinegar. It then becomes of a fine black color.

2. STAINING FINE WOODS.—Applicable especially to apple, pear, and walnut woods. Four ounces gall-nuts. one ounce powdered logwood, ½ oz. vitriol, and ½ oz. verdigris are boiled with water, and the solution, filtered hot, is applied to the wood, which is then coated with a solution of one ounce fine iron-filings dissolved in wine vinegar.

Inlaid wood is treated with a liquid prepared by mixing ½ oz. sal- ammoniac with a quantity of steel-filings, adding vinegar, and leaving the mixture for fourteen days in a warm oven. In this liquid the wood is boiled, and then allowed to remain therein for three days; it is then similarly treated with a mixture of gall-nuts and Brazil-wood decoction.

A CLEAR FLEXIBLE VARNISH.

PUSCHER recommends a solution of alum soap in oil of turpentine as a specially advantageous varnish for metallic objects which are exposed to high temperatures. Such a varnish does not possess the brilliancy of damar varnish, but it has greater flexibility and resists heat better. The alum soap is prepared by adding a solution of alum to a boiling soap solu- tion as long as a precipitate ensues, washing the latter, and drying. This alum soap is transparent, like glycerine soap, and soluble, in all propor- tions, in oil of turpentine.—*Polyt. Centralbl.*

PARISIAN WOOD-VARNISH.

TO PREPARE a good varnish for fancy woods, dissolve one part of good shellac in three to four parts of alcohol of 92 per cent in a water bath. and cautiously add distilled water until a curdy mass separates out, which is collected and pressed between linen; the liquid is filtered through paper, all the alcohol removed by distillation from the water bath, and the resin removed and dried at 100 degs. Centigrade until it ceases to lose weight; it is then dissolved in double its weight of alcohol of at least 96 per cent, and the solution perfumed with lavender oil.

CLARIFYING VARNISH, ETC.

A METHOD of clarifying liquids designed by F. Kersting, of Washington, by which it is stated that the time used in removing impurities from varnishes by settling may be reduced from several months to forty-eight hours, is as follows: Mix with every ten gallons of varnish one half-pound each of powdered marble-dust and burnt oyster-shells. All the impurities in the varnish will, it is stated, be attracted by and adhere to the oyster-shell dust, and the weight of the marble-dust mixed therewith precipitates every floating particle to the bottom of the receptacle containing the varnish. This process may be also applied to the clarification of turpentine, oils, and molasses.

PREPARATION OF ALCOHOLIC LAKES.

By A. MORELL.—Amongst the many spirit varnishes, the gold varnishes are especially defective in not adhering firmly enough to a metallic surface.

To remedy this, pure crystallized boric acid is dissolved in the varnish to the extent of half a part in 100 parts of the varnish. Thus improved, the varnish poured upon a plate forms a hard glassy coating, so firmly adherent and hard as to be impenetrable on scratching with the finger-nail.

The above proportion of boric acid must be carefully adhered to, or the varnish loses its intensity of color.

NEW PLASTIC COMPOSITION.

IT is stated by H. E. Shepard, of New Haven, Ct., that by mixing pounded fragments of mica with a proper proportion of shellac, a composition is made that, when moulded into suitable forms, will present the appearance of shell on the surface, and will be much harder and tougher than other compositions of which shellac forms the basis.

POLISH FOR FURNITURE.

EIGHT parts of white wax, two parts of rosin, and half a part of Venice turpentine, are melted at a gentle heat. The warm mass, completely melted, is poured into a stone jar, agitated, and six parts of rectified oil of turpentine are added thereto. After 24 hour , the mass having the consistency of soft butter, is ready for use. Before using the paste, the furniture should be washed with soap and water and then well dried.

INDESTRUCTIBLE PUTTY.

BOIL four pounds brown umber in seven pounds linseed oil for two hours; stir in two ounces of wax; take from the fire and mix in 5½ pounds chalk and 11 pounds white-lead, and incorporate thoroughly. The latter operation is quite essential.

MANUFACTURE OF ZINC AND LEAD PIGMENTS.

FRANKLIN OSGOOD, of New York, states that by the following process the skimmings and dross made in zincing or galvanizing may be utilized in making zinc-lead: Take the zinc-dross or skimmings, mix with coal-dust and introduce into any suitable furnace. In another furnace, charge ores of lead, or lead-dross, or a salt of lead, mixing in coal in a like manner; or the metallic materials and the coal may be mixed and charged in one furnace. The fire is then kindled. and as the heat rises, the vaporized metals pass upward and are drawn through flues by an exhaust, the mixed vapors uniting and forming a combination which may be condensed in chambers or bags in chambers adapted for that purpose, forming what is known as zinc-lead. If the zinc skimmings are "ammoniacal" they should be first deprived of alkali. Any of the known ores of zinc may be used in place of the skimmings or dross, if preferred, by first roasting the same so as to separate any acid.

MANUFACTURE OF WHITE-LEAD.

EDWARD MILNER, of Warrington, England, manufactures white-lead by the following process: Mix with finely-ground litharge a solution of the chlorides of sodium, potassium, or ammonium, and keep the mixture thoroughly agitated for a few hours. Next pass a current of carbonic acid and violently agitate the compound until it no longer gives an alkaline reaction, when the product will be found to be carbonate of lead of great body and capacity, which only needs washing (to free it from salts) and drying. to be ready for use.

DULL VARNISH.

A VARNISH which does not reflect light is prepared by mixing a solution of resin with some liquid in which resin is insoluble. A mixture of three to five parts of sandarac dissolved in forty-eight parts of ether and two and a half parts of benzol, resembles ground glass when dry. A solution of dammar resin in benzol mixed with ether also gives a good dull varnish. Water renders the varnish semi-opaque. Alfred Hughes recommends the following receipt: Ether, 560 grammes; benzol, 240 grammes; sandarac, 40 grammes; Canada balsam, 10 grammes.

VERMILION ANHYDROUS CUPROUS OXIDE.

BY R. BOTTGER.—Two parts of potassium hydrate are dissolved in 16 parts of water in a porcelain basin, one part of starch sugar is added, and then one part of cupric tartrate; the mixture is heated to about 60 deg. till it assumes a bright and deep red color, and then immediately poured into a considerable quantity of well-water previously de-aerated by boiling.

MANUFACTURE OF GLUE.

To avoid the many difficulties experienced from the changes of the atmosphere in drying glue by the usual method, O. S. Follett, of Montclair, N. J., proposes the drying of the jelly by placing it in a close box or chamber containing a water absorbent, such as chloride of calcium, etc.

MANUFACTURE OF WHITE LEAD.

INSTEAD of vaporizing the acetic acid by direct heat in the usual manner, A. P. Meyler, of New Britain, Ct., proposes passing the carbonic acid gas through the acid so as to compel a partial vaporization of the latter, and to carry over into the corroding-chamber only so much as can be economically used, which, he states, possesses important advantages over the old process in the saving of cost of apparatus, fuel, and attendance, besides producing a better quality of lead.

GRAINING.

MR. MASURY, in his book on graining, says : The best and cheapest and most convenient simple material for making grounds for light oak, maple, ash and chestnut, is pure raw Italian sienna, tinted with pure white lead, not the so-called sienna which is sold by most paint dealers under that name, but the genuine article, which can be, and should be obtained even at some cost and trouble, the said color being one of the most useful and indispensable articles in the paint shop. For maple ground, of course the smallest quantity is required, it being necessary only to change the white to the faintest suggestion of straw color. For ash, the ground should be but little darker. For light oak more of the sienna will be required, while for chestnut a decidedly yellowish tone is wanted. Care must be taken not to make the grounds too dark. Rather in the other extreme, for the reason, that there is a remedy for a too light ground, in the application of a greater quantity of graining color, as also in the glazing coat : while a ground too dark, cannot be made lighter. For dark oak, burnt Italian sienna with white will produce a far better ground than any other *single* color. The same caution must be observed, however, in obtaining this color as recommended in the case of the raw Italian sienna. The domestic so-called siennas will not prove substitutes for the genuine Italian pigments.

The ground for black walnut may be the same as for light oak with the addition of a little burnt sienna and black. No two professed grainers, perhaps, will agree as to the exact tint of color for ground work, each one having some predilection for a particular tone. These instructions being offered, not to the expert, but to the uninitiated, we do not propose to run counter to any man's prejudices, our object being as aforesaid, to simplify the matter to the last possible degree.

RULES OF THE NAVAL STORE TRADE.

The Naval Store trade have adopted the following rules :

All contracts for the purchase or sale of Spirits Turpentine shall, unless otherwise specified, be on the basis and at the rate of seven (7) pounds net weight per gallon.

Spirits Turpentine barrels and their contents shall be weighed by pounds, and have their gross weight distinctly and conspicuously marked; while the actual tare of every barrel, after being properly glued, shall be cut or indelibly branded on one head.

Rosin shall be bought and sold by the barrel of 280 pounds gross, shall be weighed by pounds, proper allowance being made for moisture and adhering dirt, and each barrel shall have its weight distinctly marked on one head.

Buyers may examine and test, at their own expense, the accurracy of weights to extent of ten (10) per cent. of any lot, and any error thus ascertained shall be corrected by reweighing the lot by another weigher, at seller's expense, or the average difference as ascertained, may by mutual agreement be made basis of settlement.

W G Jackman

DAVID HOADLEY.

We live in an age in which the moral tone of the business community is by no means high. Not to speak of flagrant acts of dishonesty, which startle and alarm us by their frequency, the general standard of what is fair and legitimate, in the dealings of man with man, is not such as to make us very proud of our Christian civilization. One need not be much of a moralist to see and deplore this. Sharpness and shrewdness, rather than perfectly straightforward dealing, too generally characterize the money-making man. People are beginning to doubt whether, after all, honesty be really the best policy — whether a slight deviation occasionally from the perfectly straight path is not necessary to enable a man to compete with his neighbor on fair and equal terms.

It is proposed to present the life of one against whose stainless probity and perfect truth no word of doubt was ever spoken, and yet who was one of our most successful business men.

David Hoadley was born at Waterbury, Connecticut, on the 13th day of February, 1806. The busy manufacturing town of to-day was then a small quiet New England village, with the industrious farmers, the white houses and red barns, and the meeting-house, with its tapering spire. His father was a man who needed a wider and more extensive field of labor, so that when the subject of this sketch was about eight years of age, he removed with his family to New Haven.

Here the boy was able to enjoy much greater educational advantages than he could have received at his native place. The next following years were spent at school, and in the quiet of a home where his life was blessed by the influence, both in

precept and example, of a mother, whom he venerated and
loved, and to whom he never occasioned a moment of sorrow or
pain.

He was carefully prepared for entering Yale College, it
being at first the desire of his parents that he should study
a profession. His mental attainments were of a high order,
and his talents promised marked success. The last year which
he spent in study was passed in Philips' Exeter Academy, at
Andover, Massachusetts, and he returned home almost if not
entirely fitted for college. But just at this time his plan was
frustrated. He was naturally of a frail constitution. Close
and unremitting application to his books had impaired his health
to such a degree that, greatly to the regret of his friends and
himself, he was compelled to relinquish the sedentary life of a
student, and to undertake some more active employment. He
then became a clerk in the drug store of Messrs. Hotchkiss &
Durant, in New Haven. This place was his training-school in
business, and there he remained until the day of his attaining
his majority in 1827. He then started for New York to seek
his fortune, with a business capital of one thousand dollars,
received from his father, and with undaunted courage and con-
scious self-reliance.

Messrs. Frisby & Ely were at that time carrying on a drug
business in the lower portion of a building standing at the
corner of Wall and Water streets, afterwards the site of the office
of the *Journal of Commerce*. This building was burned in
1835. Here the young man was received, and the firm became
Frisby, Ely & Hoadley. But this partnership was of short
duration. Scarcely more than a year elapsed before Mr. Ely
died and Mr. Frisby retired. Mr. HOADLEY, at 24, almost a boy
in years, but a man in intellectual force and vigor, was left at the
head of the house, the sole survivor.

He then associated with himself Mr. George D. Phelps, who
died about two years ago, the firm name being Hoadley &
Phelps. The same store was occupied until the year 1833,

when Mr. John W. Fowler was admitted as a partner, and the name of the concern was changed to Hoadley, Phelps & Co. The business was then removed to 142 Water street, where the firm continued for fifteen years in uninterrupted prosperity.

Few houses in the city became better known than Hoadley, Phelps & Co. No firm excelled them in mercantile credit and integrity. They did a large business for those days, perhaps the largest of any house in their line. It was also a lucrative one. Mr. HOADLEY, as the head of the house, acquired an enviable notoriety. He was the popular man of the firm, and while he was known to be careful in business negotiations, he never permitted an appeal for a worthy object to pass unnoticed.

He was emphatically a worker. It was that same nervous, active energy which showed itself in his very movements, especially in his quick decided step, which made him a successful man. His devotion to business was ardent and even enthusiastic. He was ambitious to secure and maintain the place which he so long held among business men. His industry was indefatigable ; he never lost a moment, but applied himself, with all his energy, to whatever he undertook. His days of recreation were rare, and he never failed to return to his work at the time and hour appointed. In fact business was his chief pleasure and pastime.

His perception was acute, and his judgment excellent. In matters requiring prompt determination his quick decision rarely erred. He was remarkably systematic, and the influence of his care and order was perceptible in store and office.

During Mr. HOADLEY'S life as a drug merchant he built what was, for those days, a very fine house, at the corner of Houston and Mulberry streets, then a desirable place of residence. There he lived until, towards the close of his connection with that business, he removed to West Seventeenth street, near the Fifth avenue.

About 1830 he married Miss Mary O. Hotchkiss, daughter of Russell Hotchkiss, of New Haven. She died in 1837, and he subsequently married Miss Elizabeth C. Tappan, of Pough-keepsie.

Mr. HOADLEY was a warm and efficient friend of the worthy young man of merit. He was an active member of an institution formed about 1835, called the Young Men's Society (somewhat similar in its objects to the Young Men's Christian Association of the present day), many of whose members are now among our most eminent merchants and lawyers. His partner, Mr. Phelps, was its President, and was succeeded by the Hon. Henry E. Davies. Mr. HOADLEY's sympathies never grew old, and the struggling young man obtained from him cheering advice and encouragement, and when there was need, more substantial aid.

In 1848 Mr. HOADLEY retired from the drug business, and the firm sold their stock and good will to Messrs. Schieffelin Brothers & Co. He spent a year in settling up the affairs of the old concern, and then became Vice-President of the American Exchange Bank, under that veteran financier, David Leavitt, who had early discovered his ability. But this position was not congenial to his tastes, and in 1853, declining the Presidency of the bank, he accepted that of the Panama Railroad. This office he filled with marked success, until, at the end of eighteen years, his failing health determined him to resign.

A short account of the early history of the Panama Railroad may not here be out of place. The discovery of gold in California was followed by an extraordinary exodus of fortune-seekers from the older States, "the Argonauts of 1849," as a popular writer has termed them, all anxious to reach the distant El Dorado. But the dangers and difficulties of the way were almost insurmountable. Two long ocean voyages, in crowded and unhealthy ships, were the least of the travelers' discomforts. Transit across the Isthmus was effected only by great labor and expense, canoes being used from the mouth of the Chagres River to an Indian village named Cruces, a distance of thirteen miles, from whence mules afforded the only means of carriage, along a wretched bridle path, to Panama.

It was at once considered as of the greatest importance that a railroad should be built, and a company was organized for that purpose. The first estimate for building the road was one million of dollars; but the natural difficulties were so great that when, on the 27th day of January, 1855, the last rail was laid, nearly ten times that sum had been expended. Only the undaunted energy and undoubting faith of Mr. HOADLEY and his associate directors carried the work to its completion. Tangled jungles, swamps almost impenetrable, fevers, malaria and poisonous reptiles and insects resisted every attack of the invading civilization. Scarcely was an obstacle overcome before another, equally or more formidable, arose to take its place. The number of laborers who died was greatly exaggerated, and stories with little foundation were freely circulated as to the fatal results which would surely follow the further prosecution of the work; but it is undoubtedly true that many lost their lives in the fatal and death-dealing swamps. Incessant rains moreover stayed and impeded the completion of the enterprise.

Great difficulty was found in obtaining laborers for the work. Every possible means was employed for this purpose. Canadians, Irish laborers from New York, negroes from Jamaica, emigrants directly from Cork, natives of the Isthmus and of the adjacent countries, and coolies from China were all employed. They were found to bear the hardships and the effects of the climate in the order in which they are now mentioned, the Canadians being rarely sick, and the coolies dying off in such numbers that the survivors were removed to Jamaica and other fields of labor.

At no time did the prospects of the company seem more disheartening than in the fall of 1851. The funds had become exhausted, and only on the personal credit of the directors could more means be obtained. The stock was almost worthless, and the enterprise seemed a hopeless failure. But just at this time came the intelligence that eleven hundred passengers, who had arrived from New York in two steamers, had actually been conveyed over the road, on its gravel and construction cars, for a distance of some eight or nine miles, and that that portion of the road was available for transportation. The friends of the enterprise were inspired with new hopes, the steadfast upholders of the work were relieved from the doubts and anxieties which had almost overwhelmed them, and public confidence was restored.

In July, 1852, twenty-three miles of the road were completed, being nearly one half of the entire distance, and in January, 1854, it reached Summit Ridge, thirty-seven miles from the Atlantic terminus. The work was pushed on with great vigor, until, in February, 1855, through daily trains were running from ocean to ocean.

The character of the work is admirable. It has one hundred and seventy bridges and culverts, all of them of iron, and one of them six hundred and twenty-five feet in length. The road-bed is ballasted throughout with stone to the depth of eighteen inches, and the rails are laid on ties of lignum vitae,—a wood almost indestructible even in that wet climate. Probably a more substantial railway cannot be found in the world.

Of the admirable management of the completed road it is needless to speak. Every one knows of its great success. It is enough to say that, under the efficient direction of its President, it more than repaid the wildest dreams of it originators, until its great rival carried the California trade directly across the Continent.

Mr. HOADLEY was for many years an active trustee of the Mutual Life Insurance Company of New York, and a member of its Finance Committee. Here his careful judgment and and discrimination were exceedingly valuable, and his counsel was, in perhaps every instance, followed.

Mr. HOADLEY possessed a peculiar power of inspiring personal affection. The perfect truth and sincerity of the man were always evident; his warm sympathy was ever on the surface; his kindly winning smile spoke of purity of thought and deed most difficult of attainment. Yet it was not the truth or the sympathy or the purity alone which won the hearts of those who knew him. The influence was peculiar and indescribable, yet all felt it; the presence was that of one who insensibly yet surely won your attachment without knowing it himself. Those who saw him only in business life felt a peculiar attraction—felt that he inspired something more than respect, akin to yet differing from reverence, scarcely less than love.

He was the generous dispenser of charity. No worthy object for the improvement of his fellow-men ever appealed in vain to his open-hearted liberality; wherever there was suffering there his practical sympathy went; wherever there was grief he endeavored to assuage it; wherever want existed his aim was its relief. Benevolent societies found no surer friend, charitable institutions owed much to his active, earnest co-operation.

For the last eight or nine years of his life, Mr. HOADLEY resided at Englewood, New Jersey, attending daily to his business in the city. Even after his resignation of the Presidency of the railroad, his habits of work and application were such that he was almost daily in New York as usual. He delighted in his beautiful home, with its perfection of cultivation, and the glories of the distant view melting away to the West. No man was ever more universally loved and respected than he at the place of his suburban residence.

He was not old when he died. His quick, elastic movements, his nervous energy, his admirable judgment, and his unimpaired mental powers, indicated a man whose eye was not dimmed, or natural force abated. But an insidious and fatal disease had attacked him, and when it was hardly more than suspected, it had done its work. Quietly but surely it undermined a constitution never very strong. Everything was done for him which esteem and affection could prompt, but to no purpose. On the 20th day of August, 1873, in the quiet rest of his country house, with friends and neighbors, one and all, regarding his loss as a direct personal sorrow, quietly and without pain he died.   And thus we close the record of what one who loved him called "a beautiful life, which faded away gradually, like a glorious sunset."

The large crowded church at his funeral told of the feelings with which he was regarded.   Old men came from New York to show their esteem for the character of one whose prosperous career some of them had watched from its beginning. His business associates in large numbers evidenced their respect for their energetic co-laborer.   And the residents of the village closed their stores, and suspended their daily duties, to bow in reverent grief over the remains of one whose familiar face they should never see again.

Any sketch of Mr. HOADLEY's life which did not enlarge on the Christian grace and personal excellence of the man, would fail to give any true conception of his character.   To him religion was a vital thing, entering into every duty of life, influencing every action, regulating every thought.   What would seem, when spoken of most men, to be extravagant eulogy, is in his case the mere statement of simple fact.

He would himself however have been the first to disclaim any such exalted character.   Not the least conspicuous of his virtues was humility.   Looking at himself from within, with full knowledge of unspoken thoughts, of unexecuted desires, of germs which in most men would have borne their natural fruit, he humbly saw his own imperfections and never appreciated the grandeur of his simple Christian life.   To such a scrutinizing inward gaze errors and failings must have been sadly visible, for the best of us are human.   But to those who saw him as he appeared to the world, as the active church officer, the upright man of business, the upholder of every good work, the liberal dispenser of bounty, the pure and humble man of God, to them it seems difficult to give an adequate idea of the beauty of his character.

Such men are sent as examples. Not alone in the family and in business circles is their influence felt. It goes out from them, whether they know it or not, pervading all who come within their influence, and touching all with a benediction. The moral of Mr. HOADLEY's life is not far to seek. Especially does it come home to business men, who can learn from his story that success is entirely consistent with perfect integrity; nay, more, that the truest success depends upon integrity, and cannot be attained without it. And such a lesson our business men, and especially the younger ones, will do well to study and ponder.

REVIEW

NEW YORK OPIUM MARKET

DURING 1873.

JANUARY.

THE New York opium market opened with a very heavy stock on hand, being, January 1st, 1873, 900 cases, against, January 1st, 1872, 350 cases. In Smyrna at the same time the stock on hand January 1st, 1873, was 2,150 cases, and in the interior January 1st, 1873, 650 cases. In London the stock on hand included, January 1st, 1873, 936 cases.

The heavy stock on hand in New York was due to the high prices which ranged during 1872 in our market; and, encouraged by these favorable advices, Smyrna holders of opium consigned their surplus stock to the United States. Upon good authority it was estimated that two-thirds of that heavy stock was consigned by Smyrna merchants and held for their account. Notwithstanding this heavy stock on hand, holders were very firm, offering but little from first hands.

Sales opened the new year with $6 12½, gold, for opium in bond, advancing to $6 25, gold, at which prices buyers refused to operate. On the 6th of January cable advices from Smyrna reported an advance there to 225 piasters, which was not generally credited here, and prices receded gradually to $6, gold, in bond. About the 10th of January cable advices from Smyrna reported rumors of damage to the growing crop, but these accounts had no effect whatever at first; a day or two afterwards limited sales were reported at $6 06¼ to $6 12½, gold, in bond. A lot of 10 cases which about that time was reported as sold on private terms, and for re-export, leaked out to have been made at $5 95, gold, and for shipment to London, weakened the already not too strong market down to $6, gold.

On the 13th of January several cables were received, all agreeing that prices in Smyrna were firm at 225 piasters, yet our market still remained quiet and entirely nominal at $6, gold, in bond. About the middle of January holders were forced to submit to a further reduction, and a few isolated cases were offered and taken at $5 90, gold. On the 20th a cable from Smyrna reported a decline there to 216 piasters, and holders here reduced correspondingly their rates to $5 80, gold ; yet no buyers seemed willing to lay in heavier supplies than immediate wants would demand. On the 25th it was generally reported that the speculative movement which was set on foot the day before had induced holders to withdraw their stock from the market. The result was a better demand, and some 225 cases changed hands on private terms, said to have been within $5 75 and $5 80, gold. At the former price a few cases were bought for English account and shipped to Liverpool. Manufacturers of morphine bought some 75 to 100 cases, and the balance of about 65 cases were offered at $6, gold, of which only 20 cases found takers at that figure. No doubt the balance could have been placed also but for cable advices reporting a decline in Smyrna from 216 to 212 piasters, thus closing the month at the nominal price of $5 95, gold, in bond.

FEBRUARY.

The month of February opened rather quiet, with sales of five cases at $5 87½, gold, in bond. On the 4th of February a cable from Smyrna reported the market there weak at 210 piasters. This news depressed our market, and sales were made in consequence thereof at $5 75, gold, in bond. On the 7th of February a further reduction to 200 piasters was reported, and holders here, to meet the exigencies of the case, reduced their rates to $5 62½, gold, at which rate only 15 cases found buyers. On the 10th of February a cable from Smyrna reported limited sales at 195 piasters, upon which holders offered their stock more freely at $5 62½, gold, but without finding buyers. Three days later Smyrna advices reported the market weak at 180 piasters, and with a still further downward tendency, which soon proved true, for the next day a cable reported sales at 175 piasters. Here the market declined in one day from $5 62½ to $5 25, gold, with sales of 15 cases at $5 12½, gold, prompt cash. This created an uneasy feeling, and several days passed before an explanation from Smyrna assured holders that late low prices accepted were due to the extreme stringency of the Smyrna money market, and that the crop accounts were decidedly unfavorable, owing to the severe drought then prevailing all over Turkey, and the season altogether being adverse to all growing crops. This in itself restored confidence, and the decline in our market was effectually checked, with sales at $5 37½—a slight advance. On the 20th a cable from Smyrna confirmed previous reports of damage to quoted an advance to 200 piasters. This stimulated a more active demand, and prices rapidly advanced to $5 62½, at which rate some 70 cases changed hands, closing with sales of smaller parcels at $5 75, gold. A few days later an intercepted cable from Smyrna reported a sale of 80 cases at 195 piasters, a decline, and at the end of the month our market closed somewhat easier at from $5 70 to $5 75, gold, the latter figure being quite nominal.

MARCH.

The month of March opened rather dull. Smyrna prices having still further receded to 190 piasters, buyers refused to purchase except at a concession, which was promptly met on the part of holders, who reduced their rates to $5 62½, gold, for small parcels and of broken packages, while whole cases only found takers at $5 50, gold, in bond. On the 7th of March a cable reported prices down to 180 piasters and market weak, which brought New York prices down to $5 12½, gold. The following day a further decline was reported from Smyrna, with sales at 170 piasters. Here prices receded to $5, but even at that rate we heard of no sales for four days. On the 17th a cable from Smyrna reported a speculative feeling, with liberal sales at 175 piasters. This stimulated a better demand, and we noticed sales of 40 cases at from $5 to $5 06¼, gold. About the 25th a cable reported a slight reduction in value in Smyrna, with sales at 170 piasters. Our market nevertheless did not respond, and we noted sales at $5, closing quite firm at $5, gold, in bond.

APRIL.

The month of April opened with advices from Smyrna of an advance from 170 to 174 piasters, and holders here withdrew some of their offers, which were $4 80 to $5, gold; at the former price some 70 cases and at the latter some 40 cases sold; while 10 cases brought $5 06¼, gold. On the 4th a cable from Smyrna brought reports of a short crop in prospect and sales at 175 piasters; our market therefore became firm at $5 12½, gold, at which rate 15 cases changed hands. About the middle of the month rumors from Smyrna reporting damage to the crop were freely circulated, also that prices had advanced there to 190 piasters; neither was credited, and later letter advices reported sales at 183 piasters. Our market ruled steady at from $5 12½ to $5 18¾, with sales at the former quotation. On the 20th the same rumors were again reported by cable of damage to the growing crop; but the unsettled state of our money market prevented an advance in value. Strange to remark, notwithstanding reliable cable advices of sales at 185 piasters, our market receded from 12½c to 18¾c, gold, and single cases were sold again at $5, gold. Up to the 25th the market ruled in favor of buyers, who laid in stock, although to a limited extent only, at $5, gold, 30 days, or the usual discount for cash. On that day a cable reported an advance from 185 to 190 piasters, and the month closed here with a better feeling and sales at $5 12½, gold, in bond.

MAY.

The month of May opened with a firm market and active demand; sales in Smyrna at 190 piasters and in London at 22s 6d. In our market 37 cases changed hands at $5 12½ to $5 25, gold. On the 6th of May a cable from Smyrna reported that rain was falling, it having been very dry up to that time, and that prices had slightly receded, with sales at 183 to 185 piasters. Notwithstanding the above reports, prices here took an upward instead of a downward course, and we notice sales at $5 25, gold, with not a few holders asking $5 37½, gold. On the 10th two cables were received, but both kept private, the knowing ones offering to sell again at $5 25 and $5 20. On the 13th of May advices by mail, as compared with corresponding dates sometime previously received by cable, were conflicting, so far as the crop was concerned; prices here lost their ground, and we note small transactions at $5 12½. Up to the 20th all cables were again kept secret; and holders offering their stock more freely, accepted a slight reduction; this concession cleared a lot of 40 cases, which were in somewhat anxious hands to realize, and upon being withdrawn from the market prices closed on the 25th firm at $5 25, gold. On the 27th a cable from Smyrna reported sales at 183 piasters, with sales here in round lots at $5 25, gold, in bond. On the 30th of May confirmation of previous cable advices reporting the prospects of a short crop, were received in this city and widely circulated all over the country, raising prices at the end of the month to $5 50, gold, with sales of 100 cases at from $5 25 to $5 37½, gold. On the 31st a cable from Smyrna reported sales of 200 cases at 190 piasters.

JUNE.

The month of June opened feverish. The statement of the prospect of the new crop, which at first upon good authority was estimated to yield 3,000 cases, soon was reduced to 2,500 cases. On the 2d a cable from Smyrna reported there a large speculative demand at 210 piasters, and parties here predicted on the strength of that that prices in Smyrna would reach before July 245 piasters. At from 205 to 210 piasters liberal sales were made in Smyrna, and our market immediately responded to the advance, with sales of 75 cases at $5 50 to $5 62½. Soon upon this cables from London reported there important transactions at 23s to 24s, gold, which induced holders here to ask a further advance, with sales at $5.75, gold, in bond. On the 5th of June cables from London reported sales at 24s to 25s, a furthei advance which strengthened the views of sellers heie, and we noted sales heve of ·10 cases at $5 87½, gold, in bond ; asking piic:s not in few instances $5 9) to $6, gold. On the 10th of June a cable fiom Smyrna reported transactions at 210 piasters, and closing with a fiimei feeling at 215 piasters. Upon those advices, sales were made heie at $6, gold, in bond. On the 15th cables from ,Smyrna quoted 213 piasters, and London 24s, a slight reduction, which quieted our maiket; but we noticed no reduction in value. On the 20th 2ı cases were thrown upon our market, and to clear the same $5 87½, gold, was accepted. Smyrna cable received towards nıght, and after the sale was made, ıeported a slight improve_ ment in piices theie with sạles at 215 piasteis. From that time on until the close of the month prices ruled steady with sales at $6, gold, and in bond.

JULY.,

The month of July opened with cables from Smyrna at 217 piasters, which caused a better demand and sales of 75 cases at $6 for spot, and 10 cases August delivery at $6 25, gold. On the 8th a cable from Smyrna quoted 220 piasters, although in this market sales of 65 cases were made at from $6 12½ to $6 25, gold. On the 11th another cable quoted 225 piasters, which advanced prices here from $6 25 to $6 37½, gold. On the 15th a further advance in Smyrna to 227 piasters, with sales here at $6 37½ for spot, and $6 50 for August deliveıy. On the 21st piices receded in Smyrna to 225 piasters, prices ranging heie fiom $6 37½ to $6 50, gold. The month of July closed less firm with sales of some outside lots at $6 25, although first hands were less inclined to yield to outside pressure. The quotations, although not based on actual sales, were $6 37½, bond ; $7 40, gold, and $8 40, currency, duty paid.

AUGUST·

The month of August opened without cables trom Smyrna, and prices slightly fluctuated in consequence thereof, opening with limited sales at $6 37½, and gradually declining to $6 25, gold ; and abouc the 10th a few cases changed hands at from $6 to $6 25, gold, in bond, and $8 10 to $8 12½, currency, and duty paid. On the 15th a cable fiom Smyrna reported sales at 205 piasters, and $6 12½, gold, was paid in our maiket. On the 20th of August a sale of 10 cases reported within the limit of $6 to $6 12½, gold. On the 22d cables from London reported sales there to a consider_ able extent, and for shipment to China ; but, although that tiansaction was well known and fully credited here, prices were quite nominal at $6 12½, and the market veiy quiet. On the 26th a cab'e fiom Smyrna reported sales at 210 piasters, which strengthened the· views of holders, and the asking price reached $6 25, gold. The month closed with a sale of 5 cases at $6 18¾, gold, for spot, and $6 25, gold, September delivery.

SEPTEMBER.

The month of September opened with cable advices of an advance to 215 piasters, and sales here at $6 25, gold; asking rate $6 37½, gold. The advance in Smyrna was lost by the 10th, and no sales in bond recorded, owing to the almost nominal price then ruling here; a few cases changed hands at $7 25, gold, and $8 25, currency, and duty paid. The middle of the month still found prices ruling in favor of buyers, and duty paid; opium selling in small parcels at $8 20, currency. On the 20th a few cases sold at $6 25, gold, and broken lots at $8 30, currency, closing the month without cable dispatches, and prices entirely nominal, at $6 25, gold, in bond; $7 25, gold, duty paid, and broken case lots at $8 20 to $8 25, currency.

OCTOBER.

The month of October opened with a better feeling, influenced by more favorable advices from London and Smyrna; but already on the 5th some forced sales of duty paid opium at $7, gold, duty paid, a decline of 25c, gold, in the face of better foreign advices, unsettled the market to some extent. On the 10th cables from Smyrna reported sales at 220 to 225 piasters. A material advance was immediately demanded here, but not acceded to, and although $6 50, gold, was the ruling price for large lots, small lots from jobbers' hands sold at $7 25, gold. On the 15th another cable reported sales in Smyrna at 230 piasters. Thirty cases sold for future delivery, buyers' option, at $6 30, gold, which so demoralized a few holders that for prompt cash they parted with their stock at $6, gold, although the principal holders refused to sell on the spot below $6 50, gold. About the 20th several cables were said to have been received, but found no credence, for the difference was from 236 to 230 piasters, and prices here became quite nominal in bond, varying at from $6 25 to $6 30, gold, closing the month with limited demand and unimportant sales at $6 25, gold, in bond.

NOVEMBER,

The month of November opened with cables from Smyrna at 225 piasters, yet holders here conceded a trifle, asking $6 12½, gold, in bond, without sales; jobbing lots changed hands at $8, currency, duty paid. On the 8 h cable advices reported prices quite nominal at from 215 to 220 piasters. Sales here to a very limited extent and within range of $6 12½, gold. On the 15th prices receded in Smyrna to 210 piasters. No sales in bond were reported, and quotations entirely nominal at $6 12, gold, in bond. Duty paid opium felt the pressure, and jobbers accepted $7 87½, currency, a reduction of 12½c. On the 19th a further decline to 208 piasters was reported, which produced its effect on our market the next day; duty paid declined from $7 12½ to $7, gold. No sales in bond, although several parcels were offered at $6, gold. Towards the 24th a cable from Smyrna reported the market steady with a better demand for China, but buyers not willing to pay more than 200 piasters, at which rate some few orders were filled. Here the month closed with an entire absence of all but small orders, and sales for odd parcels were made at $6, gold, and in bond.

DECEMBER.

The month of December opened dull, and values nominal at $6, gold. On the 5th cables reported that the orders for China had been filled at from 200 to 203 piasters, and that holders were again asking 206 piasters. Our market here still flagged, and even $6, gold, was not obtainable. No sales of any importance took place during the whole month, and holders soon offered their stock at $5 95, gold; still no demand, except in the jobbing way at $6 95 to $7 gold, duty paid. About the 15th a few cases sold in bond at $5 90, and soon after single cases changed hands at $5 87½, gold, in bond, and $6 87½, gold, duty paid. Jobbing and broken case lots sold in a small way at $7 85, currency, duty paid; closing the year dull and entirely nominal at $5 50, gold in bond; $7 70 currency, and Smyrna cable quoting 185 piasters.

New York Opium Quotations, 1867–1873.

January.—1867.	Bond. Gd.	Duty paid. Currency.	Gold.
5	4 25	9 05	134
12	4 25	9 07	134
19	4 31	9 10	135
26	4 25	9 05	134
February.			
2	4 12½	8 98	136
9	4 15	9 05	138
16	4 12½	9 13	139
23	4 12½	9 10	138
March.			
4	4 12½	9 20	139
11	4 10	9 00	139
18	4 10	9 00	139
25	4 12½	8 90	134½
April.			
6	4 19	9 00	13
13	4 15	8 89	133
20	4 10	8 92	135
27	4 00	8 90	137
May.			
4	4 06	8 92	136
11	4 10	9 00	136
18	4 15	9 10	136½
25	4 25	9 23	137½
31	4 30	9 35	157
June.			
3	4 50	9 55	136½
10	4 50	9 60	137
17	4 50	9 66	138
24	4 50	9 65	137⅞

July.	Bond. Gd.	Duty paid. Currency.	Gold.
1	$4 50	$9 73	139
8	4 50	9 73	138
15	4 25	9 41	139½
22	4 30	9 50	140
29	4 50	9 80	140
August.			
5	4 50	9 83	140½
12	4 37½	9 65	140⅜
19	4 40	9 72½	141¾
26	4 50	9 94	142
September.			
2	4 25	9 72	144
9	4 25	9 65	143
16	4 20	9 57½	143
23	4 20	9 57½	143
30	4 25	9 59	143¼
October.			
7	4 15	9 61	145
14	4 20	9 72½	144½
21	4 15	9 57	144¼
28	4 12½	9 50	142½
November.			
4	4 20	9 41	140½
11	4 10	9 30	140½
18	4 00	9 10	140
25	4 00	9 00	138⅛
December.			
2	4 00	8 90	137
9	4 00	8 87½	136½
16	4 10	8 75	133½
23	4 00	8 67	133½
30	4 00	8 67	133

January.—1868.	Bond. Gd.	Duty paid. Currency.	Gold.
2	4 50	9 40	134½
8	4 55	9 57	137
15	4 62½	9 72½	139
22	4 75	10 12½	140
29	4 80	10 22	140
February.			
5	4 75	10 33	142½
12	4 75	10 25	141
19	4 80	10 29	141
26	5 00	10 63	141¾
March.			
6	5 00	10 55	140¾
13	5 00	10 42	139
20	5 00	10 47	140
27	5 00	10 40	138¾
April.			
2	5 00	10 38	138½
9	5 00	10 31½	138
16	5 00	10 38	138½
23	5 10	10 50	139½
30	5 20	10 72	140
May.			
7	5 25	10 81	139½
14	5 20	10 72	139½
21	5 20	10 74	140
28	5 25	10 81	139½
June.			
4	5 15	10 70	139
11	5 00	10 48	139¾
18	4 90	10 25	140
25	4 62½	9 97	140

JULY.			
2	4 62½	10 02	140¾
9	4 70	10 19	140½
16	4 75	10 27	142
23	4 62½	10 25	143½
30	4 87½	10 57	143
AUGUST.			
6	5 00	10 87½	145
13	5 25	11 20	145
20	5 50	11 65	145
27	5 87½	12 10	144½
SEPTEMBER.			
3	6 25	12 66	144¾
10	6 25	12 60	144
17	6 25	12 62	141¼
24	6 12½	12 16	141
OCTOBER			
1	6 75	12 95	140
8	6 50	12 52	138
15	6 12½	11 81	137
22	6 15	11 80	136½
29	6 25	11 70	133¾
NOVEMBER.			
5	6 25	11 72	134
12	8 00	14 14	134¾
19	8 12½	14 30	134½
26	8 25	14 43	135
DECEMBER.			
3	8 50	14 98	136¼
10	9 12	15 75	135½
17	10 00	16 96	135¾
24	10 50	17 67	135
31	11 75	19 20	134¾

JANUARY.—1869.			
6	12 50	20 28	135¼
13	13 00	20 96	135¼
20	12 37½	20 26	136¼
27	12 25	20 09	136¼
FEBRUARY.			
3	11 50	18 93	135¼
10	11 37½	18 63	135
17	11 25	18 35	135¼
24	11 10	17 95	133
MARCH.			
3	11 00	17 82	132
10	10 80	17 45	131¼
17	10 75	17 31	131
24	10 50	17 06	131¼
31	10 75	17 45	132½
APRIL.			
7	10 75	17 45	132½
14	11 25	18 35	133½
21	11 00	18 02	133½
28	10 75	17 85	133
MAY.			
5	10 50	17 87	137½
12	9 50	16 77	139¾
19	8 00	14 25	139½
26	7 50	13 94	139⅜
JUNE.			
2	8 00	14 51	138¼
9	8 00	14 62	139¼
16	7 50	14 28	139
23	7 25	13 75	137½
30	7 25	13 35	137

JULY.			
1	7 25	13 25	137
8	7 25	13 35	137
15	7 50	13 57	135¾
22	7 75	14 00	135½
29	8 00	14 28	136
AUGUST.			
5	8 50	14 96	136
12	7 50	13 42	134¼
19	7 00	12 51	131¾
26	7 00	12 45	131
SEPTEMBER.			
2	7 00	12 75	134¼
9	6 75	12 35	135
16	6 00	11 62	136¾
23	6 00	12 15	144
30	6 00	11 45	130
OCTOBER.			
7	5 75	10 76	130½
14	5 50	10 42	130¼
21	5 75	10 75	130
28	6 75	11 97	129½
NOVEMBER.			
4	7 00	12 04	126¾
11	7 25	12 40	126½
18	7 37	12 48	126
25	8 00	13 07	124½
DECEMBER.			
2	8 00	12 86	122½
9	8 00	12 75	121½
16	8 00	12 75	121
23	7 62	12 14	120
30	7 60	12 17	120½

New York Opium Quotations, 1867–1873.—Continued.

January—1870.	Bond. G.	Duty paid. Currency.	Gold.
5	7 62	12 35	122
12	7 50	12 20	122
19	7 25	12 00	121
26	7 00	11 50	121
February.			
2	7 12½	12 15	121
9	7 37	11 94	121
16	7 50	11 99	120
23	7 87	11 95	118¼
March.			
2	8 00	12 12	111¼
9	8 15	11 85	111
16	8 25	12 00	111¼
23	8 50	12 20	111½
30	8 75	12 55	111½
Al.			
6	7 87	11 67	112½
13	7 87	11 65	112
20	7 87	11 80	113¼
27	8 00	12 12	115½
My.			
4	7 62	11 63	115
11	7 62	11 60	114¾
18	7 62	11 63	115
25	7 50	11 45	114½
June.			
1	7 50	11 35	113½
8	7 50	11 32	113¼
15	7 60	11 45	113
22	7 75	11 42	112
29	7 75	11 48	112½

July.	Bond. Gold.	Duty paid. Currency.	Gold
6	$8 37	$12 16	111⅞
13	7 75	12 35	120½
20	7 75	12 30	120
27	7 00	11 59	122
August.			
3	6 50	10 73	118
10	6 00	9 95	117
17	6 00	9 95	117
24	6 75	10 80	116¾
31	6 75	10 80	116¾
September.			
7	6 75	10 55	114
14	6 75	10 45	113
21	7 00	10 72	112⅞
28	7 00	10 67	112
October.			
5	6 37½	10 11	114
12	5 75	9 34	113¼
19	5 75	9 30	113
26	8 00	9 48	111½
November.			
2	6 12½	9 55	110¾
9	6 15	9 75	111½
16	6 25	9 83	112⅔
23	6 12½	9 62	111⅞
30	6 12½	9 62	111¾
December.			
7	6 00	9 44	111⅛
14	6 12½	9 55	110¾
21	6 12½	9 60	111
28	6 25	9 66	110½

January.—1871.	Bond. Gold.	Duty paid. Currency.	Gold.
4	6 37	8 16	110¾
11	6 25	8 02	110⅝
18	6 25	8 00	110¼
25	6 25	8 00	110½
February.			
1	6 12½	7 95	111⅜
8	6 12½	7 92	111
15	6 12½	7 92	111⅜
22	5 75	7 52	111½
March.			
1	5 50	7 21	111
8	5 40	7 15	111
15	5 25	6 94	111⅛
22	4 75	6 36	110¾
29	4 70	6 25	110¼
Al.			
5	4 62	6 20	110⅜
12	4 45	6 03	110¾
19	4 25	5 84	111
26	4 25	5 85	111¼
May.			
3	3 75	5 28	111⅛
10	4 00	5 70	111¼
17	4 37	6 02	111⅞
24	4 50	6 25	111½
31	4 62½	6 27	111⅝
June.			
7	4 25	5 90	112⅜
14	4 25	5 90	112
21	4 50	6 19	112½
28	4 50	6 15	112

July.			
5	4 62	6 36	113⅛
12	4 62	6 36	113⅛
19	4 50	6 17	112¼
26	4 00	5 61	112⅛
August.			
2	4 25	5 89	112⅛
9	4 25	5 91	112½
16	4 25	5 91	112⅝
23	4 12½	5 77	112⅝
30	4 12½	5 75	112
September.			
6	4 12½	5 78	112⅞
13	3 95	5 64	114
20	3 90	5 52½	114
27	3 87½	5 59	114¾
[illegible]			
4	3 90	5 63	115
11	4 00	5 73	114½
18	4 00	5 73	114½
25	4 00	5 61	112¼
November.			
1	4 00	5 57	111½
8	4 00	5 52	111
15	4 00	5 56	111⅛
22	3 90	5 40	110¼
29	3 90	5 43	110¾
December.			
6	3 85	5 35	110¼
13	3 70	5 13	109⅛
20	3 75	5 22	109
27	4 00	5 46	109¼

January—1872.			
3	4 00	5 45	109⅛
10	4 00	5 45	109
17	4 00	545	109⅛
24	4 00	5 45	109⅝
31	4 00	5 48	109⅝
February			
7	4 25	5 77	109⅜
14	4 25	5 77	109
21	4 30	5 81	110⅝
28	4 50	6 21	110
March.			
7	5 00	6 61	110⅛
14	5 00	6 60	110
21	4 75	6 12	110¼
28	4 60	6 17	110
April.			
4	4 70	6 28	110¼
11	4 75	6 35	110⅜
18	4 62½	6 10	111
25	4 50	6 19	112⅝
May.			
5	4 50	6 23	112¼
12	4 40	6 17	114
19	4 37	6 10	113⅝
26	4 25	5 98	113¾
June.			
6	4 25	5 99	114⅛
13	4 12½	5 85	114
20	4 40	6 12	113⅜
27	4 50	6 27	114

July			
3	4 95	6 78	114
11	5 50	7 39	114
18	5 50	7 39	114
25	5 37½	7 33	114⅝
August.			
1	5 25	7 24	115½
8	5 25	7 25	115⅝
15	5 30	7 20	115
22	5 25	6 96	113
29	5 25	6 85	112
September.			
5	5 00	6 73	113
12	5 20	7 00	113⅛
19	5 37½	7 32	115
26	5 62½	7 36	113
October.			
2	5 62½	7 45	114
9	5 75	7 62	113
16	6 00	7 91	112⅛
25	6 25	8 20	113
November.			
6	6 25	8 29	113
13	6 25	8 35	114
20	6 00	7 91	113¼
27	6 00	7 90	113
December.			
4	6 00	7 87	112⅝
11	5 90	7 79	113
18	6 00	7 84	112
27	6 12½	7 95	113

SMYRNA OPIUM QUOTATIONS, 1850–1873.

January.—1873.	Bond. ℔.	Duty paid. Currency.	Gold.
2	6 12½	8 00	112
8	6 00	7 84	112
15	6 00	7 84	112½
22	5 80	7 75	113½
29	5 87½	7 75	114
February.			
5	5 87½	7 80	113¾
12	5 50	7 50	114
19	5 37½	7 40	114½
26	5 62½	7 80	115
March.			
5	5 50	7 50	115
12	5 00	7 00	115
19	5 12½	7 10	115¼
26	5 00	7 00	116
April.			
2	5 00	7 10	117
9	5 12½	7 37½	118
16	5 25	7 40	117½
23	5 00	7 10	117¼
30	5 00	8 10	117
May.			
7	5 25	7 37½	117½
14	5 12½	7 25	118
21	5 06¼	7 20	117½
28	5 25	7 50	116
June.			
4	5 75	8 30	118
11	8 00	8 32½	117½
18	5 90	8 20	116
25	6 00	8 31½	115½

July.	Bond. ℔.	Duty pd. Currency.	Gold
2	6 00	8 20	115
9	6 12½	8 30	115
16	6 25	8 45	116
23	6 50	8 70	115½
30	6 37½	8 58	115½
August.			
6	6 37½	8 50	115½
13	6 25	8 40	115½
20	6 12½	8 25	115½
27	6 18¾	8 20	115
September.			
3	6 25	8 40	116
10	6 25	8 10	111
17	6 25	8 10	111
24	6 25	8 12½	112
October.			
1	6 25	8 10	111
8	6 25	8 00	110
15	6 37½	8 05	109
22	6 30	7 95	108½
29	6 35	7 90	108½
November.			
5	6 12½	7 65	107
12	6 12½	7 65	107
19	6 00	7 70	109
26	6 00	7 70	109
December.			
3	6 00	7 70	109
10	6 00	7 75	109½
17	5 95	7 85	112
24	5 87½	7 85	110
31	5 50	7 70	110½

SMYRNA.

January—1850.		
2	Piasters	94
9	"	93
16	"	93
23	"	92
30	"	91
February.		
6	Piasters	91
13	"	92
20	"	92
27	"	94
March.		
6	Piasters	95
13	"	97
20	"	98
27	"	100
April.		
3	Piasters	102
10	"	104
17	"	102
24	"	100
May.		
1	Piasters	95
8	"	91
15	"	90
22	"	85
29	"	81
June.		
5	Piasters	79
12	"	81
19	"	82
26	"	83

JULY.

Date		Piasters
5	.....Piasters	93
12	"	93
19	"	94
26	"	95

AUGUST.

Date		Piasters
2	.....Piasters	94
9	"	93
16	"	92
23	"	90
30	"	89

SEPTEMBER.

Date		Piasters
6	.....Piasters	88
13	"	88
20	"	90
27	"	93

OCTOBER.

Date		Piasters
4	.....Piasters	90
11	"	89
18	"	89
25	"	88

NOVEMBER.

Date		Piasters
1	.....Piasters	87
8	"	85
15	"	88
22	"	88
29	"	90

DECEMBER.

Date		Piasters
6	.....Piasters	90
13	"	92
20	"	92
27	"	91

JANUARY.—1851.

Date		Piasters
4	....Piasters	90
11	"	89
18	"	87
25	"	85

FEBRUARY.

Date		Piasters
1	.....Piasters	84
8	"	83
15	"	82
22	"	81

MARCH.

Date		Piasters
1	.....Piasters	79
8	"	79
15	"	78
22	"	77
29	"	78

APRIL.

Date		Piasters
5	.....Piasters	77
12	"	79
19	"	85
26	"	87

MAY.

Date		Piasters
3	.....Piasters	90
10	"	90
17	"	89
24	"	91
31	"	92

JUNE.

Date		Piasters
7	.....Piasters	91
14	"	92
21	"	92
28	"	93

JULY.

Date		Piasters
8	.....Piasters	85
10	"	84
17	"	86
24	"	85
31		86

AUGUST.

Date		Piasters
7	.....Piasters	87
14	"	87
21	"	86
28	"	85

SEPTEMBER.

Date		Piasters
4	.....Piasters	84
11	"	85
18	"	87
25	"	89

OCTOBER.

Date		Piasters
2	.....Piasters	90
9	"	91
16	"	92
23	"	92
30	"	93

NOVEMBER.

Date		Piasters
6	.....Piasters	94
13	"	94
20	"	93
27	"	92

DECEMBER.

Date		Piasters
4	.....Piasters	91
11	"	92
18	"	92
27	"	91

Highest price paid during 1850. ..104 piasters, N. C.

Lowest " " " 79 " "

SMYRNA OPIUM QUOTATIONS, 1850-1873 CONTINUED.

JANUARY.—1852. — Piasters

Date	Piasters
3	92
10	95
17	97
24	99
31	99

FEBRUARY.

Date	Piasters
7	98
14	97
21	93
28	99

MARCH.

Date	Piasters
6	98
13	95
20	96
27	97

APRIL.

Date	Piasters
3	97
10	95
17	95
24	96

MAY.

Date	Piasters
1	92
8	91
15	91
22	89
29	88

JUNE.

Date	Piasters
5	85
12	84
19	84
26	88

JULY. — Piasters

Date	Piasters
3	85
10	88
17	93
24	95
31	98

AUGUST.

Date	Piasters
7	95
14	90
21	87
28	83

SEPTEMBER.

Date	Piasters
4	83
11	84
18	83
25	84

OCTOBER.

Date	Piasters
2	85
9	83
16	83
23	82
30	81

NOVEMBER.

Date	Piasters
6	83
13	85
20	83
27	84

DECEMBER.

Date	Piasters
4	85
11	85
18	87
27	87

Highest price paid during 1852....99 piasters, N. C.
Lowest   "   "   "   81

JANUARY.—1853. — Piasters

Date	Piasters
3	87
8	85
15	86
22	85
29	84

FEBRUARY.

Date	Piasters
5	87
12	89
19	92
26	90

MARCH.

Date	Piasters
5	88
12	89
19	87
26	86

APRIL.

Date	Piasters
2	86
9	85
16	87
23	88
30	89

MAY.

Date	Piasters
7	88
14	87
21	89
28	90

JUNE.

Date	Piasters
4	89
11	88
18	87
25	87

1853

JULY.
	Piasters
2	86
9	86
16	87
23	88
30	89

AUGUST.
	Piasters
6	90
13	90
20	91
27	91

SEPTEMBER.
	Piasters
3	90
10	90
17	91
24	91

OCTOBER.
	Piasters
1	90
8	89
15	89
22	90
29	91

NOVEMBER.
	Piasters
5	89
12	90
19	89
26	88

DECEMBER.
	Piasters
3	86
10	87
17	88
24	90

Highest price paid during 1853....92 piasters, N.C.
Lowest " " " " " 84

1854

JANUARY—1854.
	Piasters
7	91
14	88
21	95
28	91

FEBRUARY.
	Piasters
4	91
11	91
18	90
25	91

MARCH.
	Piasters
4	92
11	91
18	92
25	90

APRIL.
	Piasters
1	90
8	91
15	90
22	90
29	92

MAY.
	Piasters
6	92
13	91
20	93
27	96

JUNE.
	Piasters
3	95
10	97
17	98
24	99

JULY.
	Piasters
1	100
8	102
15	101
22	99
29	107

AUGUST.
	Piasters
5	95
12	93
19	104
26	106

SEPTEMBER.
	Piasters
3	108
10	111
17	113
24	116

OCTOBER.
	Piasters
1	120
8	121
15	123
22	124
29	126

NOVEMBER.
	Piasters
6	125
13	127
20	128
27	129

DECEMBER.
	Piasters
4	130
11	132
18	134
27	135

Highest price paid during 1854...135 piasters, N.C.
Lowest " " " " 87

SMYRNA OPIUM QUOTATIONS, 1850–1873 CONTINUED.

JANUARY—1855.

Date	Piasters
6	139
13	137
20	138
27	137

FEBRUARY.

Date	Piasters
3	137
10	136
17	134
24	130

MARCH.

Date	Piasters
8	129
10	127
17	124
24	123
31	120

APRIL.

Date	Piasters
7	120
14	119
21	117
28	120

MAY.

Date	Piasters
5	120
12	121
19	121
25	119

JUNE.

Date	Piasters
2	111
9	105
16	100
23	93
30	92

JULY.

Date	Piasters
7	91
14	93
21	92
28	93

AUGUST.

Date	Piasters
4	93
11	92
18	91
25	91

SEPTEMBER.

Date	Piasters
1	92
8	91
15	93
22	94
29	93

OCTOBER.

Date	Piasters
6	94
13	93
20	92
27	95

NOVEMBER.

Date	Piasters
3	96
10	97
17	99
24	103

DECEMBER.

Date	Piasters
1	105
8	108
15	105
22	105
29	104

Highest price paid during 1855.. 139 piasters, N. C.
Lowest " " " 91

JANUARY—1856.

Date	Piasters
5	104
12	103
19	103
26	102

FEBRUARY.

Date	Piasters
2	101
9	101
16	100
23	99

MARCH.

Date	Piasters
3	97
10	98
17	98
24	99
31	100

APRIL.

Date	Piasters
6	101
13	102
20	104
27	105

MAY.

Date	Piasters
5	106
12	106
19	107
26	108

JUNE.

Date	Piasters
2	110
9	111
16	112
23	116
30	118

JANUARY.—1857.

	Piasters
3	124
10	125
17	123
24	122
31	121

FEBRUARY.

	Piasters
7	122
14	120
21	121
28	118

MARCH.

	Piasters
7	115
14	116
21	113
28	111

APRIL.

	Piasters
4	110
11	109
18	107
25	106

MAY.

	Piasters
2	107
9	108
16	109
23	110
30	108

JUNE.

	Piasters
6	107
13	105
20	100
27	96

JULY.

	Piasters
4	97
11	99
18	100
25	102

AUGUST.

	Piasters
1	103
8	105
15	104
22	106
29	108

SEPTEMBER.

	Piasters
5	115
12	120
19	128
26	139

OCTOBER.

	Piasters
3	137
10	138
17	139
24	135
31	130

NOVEMBER.

	Piasters
7	131
14	129
21	122
28	120

DECEMBER.

	Piasters
5	121
12	119
19	122
28	123

Highest price paid during 1857 ... 139 piasters, N. C.
Lowest " " 96

JULY.

	Piasters
6	120
13	124
20	127
27	129

AUGUST.

	Piasters
4	130
11	131
18	132
25	133

SEPTEMBER.

	Piasters
1	131
8	133
15	126
22	130
29	128

OCTOBER.

	Piasters
6	128
13	127
20	126
27	125

NOVEMBER.

	Piasters
3	125
10	123
17	125
24	124

DECEMBER.

	Piasters
1	124
8	125
15	124
22	123
29	124

Highest price paid during 1850. ..133 piasters, N. C.
Lowest " " 97

SMYRNA OPIUM QUOTATIONS, 1850-1873 CONTINUED.

JANUARY—1858.

2	Piasters	123
9	"	122
16	"	124
9	"	125
30	"	126

FEBRUARY.

6	Piasters	125
13	"	125
20	"	127
27	"	125

MARCH.

6	Piasters	123
18	"	122
20	"	121
27	"	141

APRIL.

3	Piasters	126
10	"	127
17	"	128
2	"	129

MAY.

1	Piasters	130
8	"	132
15	"	131
22	"	132
29	"	135

JUNE.

5	Piasters	131
2	"	129
19	"	127
26	"	129

JULY.

3	Piasters	130
10	"	132
17	"	131
24	"	133
31	"	135

AUGUST.

7	Piasters	139
14	"	141
21	"	143
28	"	145

SEPTEMBER.

4	Piasters	150
11	"	151
18	"	154
25	"	157

OCTOBER.

2	Piasters	156
9	"	154
16	"	152
23	"	150
30	"	151

NOVEMBER.

6	Piasters	152
13	"	150
20	"	151
27	"	154

DECEMBER.

4	Piasters	153
11	"	155
18	"	156
25	"	158

Highest price paid during 1858. 157 piasters, N. C
Lowest   "   "   "   121

JANUARY—1859.

3	Piasters	39
10	"	60
15	"	64
22	"	66
29	"	67

FEBRUARY.

5	Piasters	69
12	"	18
19	"	170
26	"	171

MARCH.

5	Piasters	170
12	"	172
19	"	171
26	"	172

APRIL.

2	Piasters	170
9	"	168
16	"	166
23	"	62
30	"	38

MAY.

7	Piasters	150
14	"	140
21	"	121
28	"	124

JUNE.

4	Piasters	128
11	"	131
18	"	136
25	"	139

JANUARY—1860.

Date	Piasters
7	152
14	150
21	153
28	156

FEBRUARY.

Date	Piasters
4	158
11	161
18	164
25	166

MARCH.

Date	Piasters
4	168
11	169
18	171
25	172

APRIL.

Date	Piasters
1	177
8	183
15	161
22	180
29	1:2

MAY.

Date	Piasters
6	160
13	152
20	141
27	140

JUNE.

Date	Piasters
3	138
10	136
17	134
24	132

JULY.

Date	Piasters
1	136
8	139
15	144
22	147
29	142

AUGUST.

Date	Piasters
5	141
12	142
19	139
26	140

SEPTEMBER.

Date	Piasters
4	142
11	141
18	140
25	140

OCTOBER.

Date	Piasters
7	141
14	141
21	140
28	141

NOVEMBER.

Date	Piasters
4	141
11	140
18	142
25	141

DECEMBER.

Date	Piasters
4	142
11	141
18	140
25	139

Highest price paid during 1860....183 piasters. N.C.
Lowest "   "   "   ....132 "   "

JULY.

Date	Piasters
2	137
9	135
16	132
23	129
30	131

AUGUST.

Date	Piasters
6	133
13	132
20	134
27	136

SEPTEMBER.

Date	Piasters
3	138
10	140
17	142
24	140

OCTOBER.

Date	Piasters
1	140
8	138
15	137
22	136
29	138

NOVEMBER.

Date	Piasters
5	141
12	144
19	117
26	149

DECEMBER.

Date	Piasters
3	150
10	148
17	149
27	151

Highest price paid during 1859...172 piasters, N.C.
Lowest "   "   "   ...121 "

SMYRNA OPIUM QUOTATIONS, 1850–1873 [CONTINUED.]

JANUARY—1861.

5.....	Piasters	139
12.....	"	138
19.....	"	139
26.....	"	137

FEBRUARY.

2.....	Piasters	136
9.....	"	136
16.....	"	134
23.....	"	135

MARCH.

4.....	Piasters	136
11.....	"	135
18.....	"	136
25.....	"	134

APRIL.

6.....	Piasters	138
13.....	"	139
20.....	"	140
27.....	"	142

MAY.

4.....	Piasters	144
11.....	"	145
18.....	"	146
25.....	"	147
31.....	"	145

JUNE.

3.....	Piasters	140
10.....	"	137
17.....	"	135
24.....	"	132

JULY.

6.....	Piasters	131
13.....	"	129
20.....	"	127
27.....	"	128

AUGUST.

3.....	Piasters	121
10.....	"	120
17.....	"	119
24.....	"	117
31.....	"	118

SEPTEMBER.

3.....	Piasters	117
10.....	"	119
17.....	"	118
24.....	"	119

OCTOBER.

5.....	Piasters	117
12.....	"	118
19.....	"	119
26.....	"	117

NOVEMBER.

2.....	Piasters	116
9.....	"	117
16.....	"	115
23.....	"	113
30.....	"	111

DECEMBER.

2.....	Piasters	112
9.....	"	114
16.....	"	116
27.....	"	115

Highest price paid during 1861 . . 147 plasters, N. C.
Lowest   "   "   "   " 111

JANUARY.—1862.

4.....	Piasters	115
11.....	"	116
18.....	"	115
25.....	"	114

FEBRUARY.

1.....	Piasters	113
8.....	"	112
15.....	"	113
22.....	"	114
29.....	"	112

MARCH.

4.....	Piasters	111
11.....	"	109
18.....	"	110
25.....	"	108

APRIL.

5.....	Piasters	109
12.....	"	110
19.....	"	108
26.....	"	111

MAY.

3.....	Piasters	114
10.....	"	115
17.....	"	118
24.....	"	120
31.....	"	124

JUNE.

3.....	Piasters	127
10.....	"	128
17.....	"	129
24.....	"	131

JULY.

5	Piasters	135
12	"	142
19	"	151
26	"	153

AUGUST.

2	Piasters	155
9	"	156
16	"	157
23	"	155
30	"	152

SEPTEMBER.

6	Piasters	150
13	"	141
20	"	142
27	"	144

OCTOBER.

4	Piasters	146
11	"	142
18	"	144
25	"	145

NOVEMBER.

1	Piasters	147
8	"	149
15	"	152
22	"	151
29	"	153

DECEMBER.

6	Piasters	155
13	"	155
20	"	156
27	"	156

Highest price paid during 1862 .. 157 piasters, N. C.
Lowest " " 108 "

JANUARY.—1863.

3	Piasters	155
10	"	155
17	"	156
24	"	155
31	"	154

FEBRUARY.

7	Piasters	155
14	"	154
21	"	157
28	"	155

MARCH.

7	Piasters	155
14	"	154
21	"	154
28	"	155

APRIL.

4	Piasters	154
11	"	152
18	"	154
25	"	153

MAY.

4	Piasters	150
11	"	149
18	"	147
25	"	146
31	"	144

JUNE.

6	Piasters	141
13	"	140
20	"	139
27	"	138

JULY.

3	Piasters	139
10	"	137
17	"	139
24	"	138
31	"	140

AUGUST.

1	Piasters	142
8	"	147
15	"	150
22	"	149
29	"	147

SEPTEMBER.

5	Piasters	148
12	"	147
19	"	146
26	"	145

OCTOBER.

3	Piasters	146
10	"	147
17	"	145
24	"	146
31	"	144

NOVEMBER.

7	Piasters	145
14	"	144
21	"	143
28	"	142

DECEMBER.

5	Piasters	144
12	"	143
19	"	145
26	"	146

Highest price paid during 1863 .. 157 piasters, N. C.
Lowest " " 142 "

SMYRNA OPIUM QUOTATIONS, 1850–1873 CONTINUED.

JANUARY—1864.
Date		Piasters
2	Piasters	144
9	"	145
16	"	147
23	"	147
30	"	148

FEBRUARY.
7	Piasters	151
14	"	153
21	"	153
23	"	154

MARCH.
7	Piasters	154
14	"	155
21	"	154
28	"	153

APRIL.
4	Piasters	152
11	"	152
18	"	151
25	"	150

MAY.
2	Piasters	151
9	"	151
16	"	152
23	"	150
30	"	119

JUNE.
6	Piasters	149
13	"	148
20	"	146
27	"	143

JULY.
4	Piasters	142
11	"	141
18	"	141
25	"	140

AUGUST.
1	Piasters	141
8	"	142
15	"	141
22	"	140
29	"	137

SEPTEMBER.
5	Piasters	136
12	"	135
19	"	133
26	"	130

OCTOBER.
3	Piasters	128
10	"	126
17	"	123
24	"	120
31	"	123

NOVEMBER.
7	Piasters	126
14	"	127
21	"	130
28	"	128

DECEMBER.
5	Piasters	25
12	"	123
19	"	22
26	"	122

Highest price paid during 1864. 155 piasters, N. C.
Lowest " " " 120 " "

JANUARY—1865.
7	Piasters	123
14	"	122
21	"	123
28	"	124

FEBRUARY.
4	Piasters	125
11	"	125
18	"	126
25	"	124

MARCH.
4	Piasters	123
11	"	122
18	"	121
25	"	119

APRIL.
1	Piasters	117
8	"	116
15	"	113
22	"	111
29	"	109

MAY.
6	Piasters	110
13	"	110
20	"	111
27	"	09

JUNE.
3	Piasters	110
10	"	111
17	"	109
24	"	107

JANUARY—1866.

JULY.

		Piasters
7		132
14	"	133
21	"	135
28	"	137

AUGUST.

		Piasters
4		141
11	"	148
18	"	155
25	"	154

SEPTEMBER.

		Piasters
1		150
8	"	148
15	"	146
22	"	146
29	"	147

OCTOBER.

		Piasters
6		147
13	"	146
20	"	146
27	"	146

NOVEMBER.

		Piasters
3		145
10	"	145
17	"	145
24	"	144

DECEMBER.

		Piasters
1		142
8	"	141
15	"	140
22	"	142
29	"	140

Highest price paid during 1866 . . . 155 piasters, N. C.
Lowest " " . . .113 "

JANUARY—1866.

		Piasters
6		115
13	"	114
20	"	115
27	"	113

FEBRUARY.

		Piasters
3		114
10	"	113
17	"	116
24	"	118

MARCH.

		Piasters
3		125
10	"	128
17	"	132
24	"	136
31	"	134

APRIL.

		Piasters
7		133
14	"	132
21	"	130
28	"	128

MAY.

		Piasters
5		125
12	"	130
19	"	121
26	"	123

JUNE.

		Piasters
2		124
9	"	126
16	"	127
23	"	129
30	"	131

JULY.

		Piasters
1		105
8	"	100
15	"	95
22	"	93
29	"	96

AUGUST.

		Piasters
5		98
12	"	99
19	"	100
26	"	102

SEPTEMBER.

		Piasters
4		102
11	"	103
18	"	102
25	"	104

OCTOBER.

		Piasters
7		104
14	"	105
21	"	106
28	"	106

NOVEMBER.

		Piasters
4		107
11	"	108
18	"	109
25	"	110

DECEMBER.

		Piasters
4		110
11	"	110
18	"	111
27	"	113

Highest price paid during 1865 . . .136 piasters, N.C.
" " " . . . 93 "

SMYRNA OPIUM QUOTATIONS, 1850–1873 CONTINUED.

JANUARY.—1867.

5	.....Piasters	141
12	.. "	140
19	.. "	142
26	.. "	144

FEBRUARY.

2	.....Piasters	144
9	.. "	143
16	.. "	142
23	.. "	140

Mr.

4	.....Piasters	140
11	.. "	138
18	.. "	135
25	.. "	131

APRIL.

6	.....Piasters	129
13	.. "	127
20	.. "	126
27	.. "	124

MAY.

4	.....Piasters	123
11	.. "	121
18	.. "	119
25	.. "	118
31	.. "	119

JUNE.

3	.....Piasters	120
10	.. "	119
17	.. "	118
24	.. "	120

JULY.

6	.....Piasters	123
13	.. "	120
20	.. "	143
27	.. "	140

AUGUST.

3	.....Piasters	135
10	.. "	130
17	.. "	133
24	.. "	132
31	.. "	131

SEPTEMBER.

3	.....Piasters	130
10	.. "	129
17	.. "	129
24	.. "	126

OCTOBER.

5	.....Piasters	129
12	.. "	128
19	.. "	129
26	.. "	128

NOVEMBER.

2	.....Piasters	128
9	.. "	128
16	.. "	127
23	.. "	128
30	.. "	127

DECEMBER.

2	.....Piasters	126
9	.. "	132
16	.. "	136
23	.. "	139

Highest price paid during 1867....114 piasters, N. C.
Lowest   "   "   "   118   "

JANUARY.—1868.

4	..Piasters	161
11	.. "	160
18	.. "	161
25	.. "	162

FEBRUARY.

1	..Piasters	158
8	.. "	155
15	.. "	152
22	.. "	150
29	.. "	160

MARCH.

6	..Piasters	163
13	.. "	159
20	.. "	157
27	.. "	155

APRIL.

5	..Piasters	156
12	.. "	160
19	.. "	165
26	.. "	170

MAY.

3	..Piasters	165
0	.. "	157
7	.. "	150
24	.. "	152
31	.. "	145

JUNE.

7	..Piasters	150
14	.. "	149
21	.. "	148
28	.. "	147

JULY.

5	Piasters	145
12	"	149
19	"	152
26	"	155

AUGUST.

2	Piasters	180
9	"	181
16	"	182
23	"	180
30	"	188

SEPTEMBER.

6	Piasters	195
13	"	205
20	"	212
27	"	205

OCTOBER.

4	Piasters	200
11	"	201
18	"	195
25	"	198

NOVEMBER.

1	Piasters	205
8	"	220
15	"	260
22	"	268
29	"	275

DECEMBER.

6	Piasters	280
13	"	300
20	"	350
27	"	355

Highest price paid during 1868 ...355 piasters, N. C.
Lowest   "   "   ...145

JANUARY—1869.

2	Piasters	360
9	"	370
16	"	375
23	"	380
30	"	290

FEBRUARY.

6	Piasters	380
13	"	372
20	"	363
27	"	360

MARCH.

6	Piasters	363
13	"	363
20	"	350
27	"	352

APRIL.

3	Piasters	295
10	"	296
17	"	295
24	"	294

MAY.

1	Piasters	291
8	"	295
15	"	300
22	"	295
29	"	282

JUNE.

5	Piasters	275
12	"	268
19	"	265
26	"	255

JULY.

3	Piasters	242
10	"	238
17	"	225
24	"	246
31	"	255

AUGUST.

7	Piasters	290
14	"	275
21	"	240
28	"	225

SEPTEMBER.

4	Piasters	200
11	"	202
18	"	200
25	"	190

OCTOBER.

2	Piasters	200
9	"	201
16	"	190
23	"	222
30	"	230

NOVEMBER.

6	Piasters	224
13	"	229
20	"	235
27	"	270

DECEMBER.

4	Piasters	255
11	"	260
18	"	264
27	"	270

Highest price paid during 1869. 390 piasters, N. C.
Lowest   "   "   190

SMYRNA OPIUM QUOTATIONS, 1850–1873 CONTINUED.

JANUARY—1870.

7	Piasters	260
14	"	255
21	"	248
28	"	243

FEBRUARY.

4	Piasters	242
11	"	240
18	"	248
25	"	252

MARCH.

4	Piasters	260
11	"	265
18	"	278
25	"	285

APRIL.

1	Piasters	280
8	"	277
15	"	282
22	"	284
29	"	285

MAY.

6	Piasters	285
13	"	279
20	"	270
27	"	263

JUNE.

3	Piasters	265
10	"	265
17	"	260
24	"	255

JULY.

1	Piasters	258
8	"	265
15	"	245
22	"	210
29	"	210

AUGUST.

5	Piasters	200
12	"	190
19	"	185
26	"	230

SEPTEMBER.

4	Piasters	231
11	"	230
18	"	232
25	"	230

OCTOBER.

7	Piasters	230
14	"	210
21	"	220
28	"	205

NOVEMBER.

4	Piasters	215
11	"	212
18	"	214
25	"	212

DECEMBER.

4	Piasters	260
11	"	205
18	"	210
27	"	210

Highest price paid during 1870...285 piasters, N.C.
"   "   "   "   ...185" "

JANUARY—1871.

6	Piasters	215
13	"	212
20	"	215
27	"	200

FEBRUARY.

3	Piasters	202
10	"	200
17	"	190
24	"	192

MARCH.

3	Piasters	174
10	"	175
17	"	170
24	"	165
31	"	163

APRIL.

7	Piasters	162
14	"	155
21	"	135
28	"	130

MAY.

5	Piasters	133
12	"	135
19	"	140
26	"	150

JUNE.

2	Piasters	150
9	"	145
16	"	150
23	"	160
30	"	158

1871

JULY.

Date	Piasters
7	150
14	160
21	145
28	135

AUGUST.

Date	Piasters
4	132
11	137
18	142
25	150

SEPTEMBER.

Date	Piasters
1	147
8	140
15	138
22	140
29	137

OCTOBER.

Date	Piasters
6	137
13	145
20	142
27	139

NOVEMBER.

Date	Piasters
3	140
10	140
17	139
24	138

DECEMBER.

Date	Piasters
1	138
8	135
15	132
22	140
29	138

Highest price paid during 1871 ..215 piasters, N. C.
Lowest " " 130

1872

JANUARY.—1872.

Date	Piasters
2	140
9	138
16	138
23	137
30	136

FEBRUARY.

Date	Piasters
6	139
13	140
20	140
27	145

MARCH.

Date	Piasters
6	163
13	162
20	160
29	160

APRIL.

Date	Piasters
3	159
10	158
17	160
24	155

MAY.

Date	Piasters
1	153
8	154
15	152
22	150
29	147

JUNE.

Date	Piasters
5	146
12	145
19	155
26	170

JULY.

Date	Piasters
3	190
10	195
17	195
24	180
31	180

AUGUST.

Date	Piasters
7	190
14	185
21	180
28	180

SEPTEMBER.

Date	Piasters
4	170
11	180
18	190
25	200

OCTOBER.

Date	Piasters
2	200
9	230
16	235
23	225
30	225

NOVEMBER.

Date	Piasters
6	235
13	225
20	220
27	220

DECEMBER.

Date	Piasters
4	210
11	210
18	220
25	225

Highest price paid during 1872 ..235 piasters, N. C.
Lowest " " " 136

Smyrna Opium Quotations, 1850–1873 Continued.

JANUARY.—1873.		
2	Piasters	220
8	"	225
15	"	220
22	"	216
29	"	212

FEBRUARY.		
5	Piasters	200
12	"	195
19	"	175
26	"	200

MARCH.		
5	Piasters	190
12	"	170
19	"	175
26	"	170

APRIL.		
2	Piasters	174
9	"	180
16	"	183
22	"	185
29	"	190

MAY.		
7	Piasters	185
14	"	180
21	"	175
28	"	183

JUNE.		
4	Piasters	2.0
11	"	215
18	"	213
25	"	215

JULY.		
2	Piasters	217
9	"	220
16	"	225
23	"	227
30	"	225

AUGUST.		
6	Piasters	220
13	"	210
20	"	210
27	"	210

SEPTEMBER.		
3	Piasters	215
10	"	210
17	"	210
24	"	210

OCTOBER.		
1	Piasters	215
8	"	220
15	"	225
22	"	230
29	"	225

NOVEMBER.		
5	Piasters	225
12	"	215
19	"	2.8
26	"	200

DECEMBER.		
3	Piasters	205
10	"	205
17	"	200
24	"	195
31	"	185

Highest price paid during 1873...230 piasters, N. C.
Lowest " " " ...110 "

DRUGS.

ON THE MANUFACTURE OF ETHER.

O. SUFFENGUTH states that the best method of making large quantities of ether is by the continuous process. A retort, containing a mixture of nine parts sulphuric acid, of 66 degs. B. and five parts 90 per cent alcohol, is heated to 284 degs. Fahr., and alcohol allowed to flow in continuously, to keep the mixture at a constant level. Heretofore a direct fire has been applied under the copper or iron retort; but, owing to the inflammability and volatility of the ether, this is evidently dangerous; and, moreover, the direct fire soon destroys the retort. or at least dissolves the leaden lining. This is now entirely avoided by the use of superheated high-pressure steam for heating the retort. Even though this method is rather more expensive, it prevents igniting and exploding the ether vapor, which quite compensates for the cost. Another advantage is the ease with which a constant temperature is maintained, by regulating the pressure, so that the operation is no longer dependent upon the care and experience of the workmen.

Various materials have been used for the retort or still; sometimes copper alone, sometimes copper lined with lead, and also iron lined with lead. Experience has proved that the last-named is not only the cheapest, but will last the longest. If the operation is carefully conducted, 66 per cent of ether, of a specific gravity of 0·730, will be obtained. Half a pound of sulphuric acid makes 100 pounds of ether, and the apparatus is so constructed that it can be refilled without interrupting the operation. Great attention to the regulation of the temperature and to the flowing in of the alcohol are the principal conditions for obtaining a large yield.

The crude ether thus obtained is freed from the acid dissolved in it, and washed, after which it is rectified in a suitable apparatus. Attempts have been made to rectify it in the process of its manufacture, by conducting the ether-vapor into a vessel with double walls. the space between the walls being filled with water at a temperature of 35 deg. C., (95 deg. Fahr.) Here the water and alcohol are condensed; while the ether passes up into a second vessel filled with pieces of quicklime, of the size of a man's fist, which take up the sulphurous acid. It is now warmed, and enters from beneath into a cylinder holding a leaden basket of dried wood-charcoal, or alternate layers of charcoal and pieces of coke soaked in a solution of soda and well dried. From here it is conducted, through a cooler, into the receiver. This continuous rectification is more difficult, and requires greater attention on the part of the workmen than where the purification is a separate operation : first, on account of the continual regulation of the temperature in the different parts of the apparatus ; and secondly, because the lime sometimes stops up the tube, or is carried off in the vapor. The operation never goes on regularly, nor is the product always pure. It seems to be better, in practice, to keep separate the two operations of making and of purifying the ether.

THE MANUFACTURE OF MAGNESIA.

THE Washington factory, near Newcastle, England, manufactures the greater part of the magnesia used in the world. The principle of the process employed consists in treating dolomite with gaseous carbonic acid, under a pressure of five or six atmospheres. The dolomite is first dried, then finely pulverized, and afterward placed, with cold water, in a cylinder which constantly revolves on its horizontal axis. The carbonic acid gas, formed by the action of hydrochloric acid upon carbonate of lime, is, by a powerful pump, driven into the vessel at the pressure above noted. The solution of bicarbonate of magnesia thus produced is carried into a vertical cylinder, and submitted to steam (the consequent elevation of temperature regenerating the neutral carbonate), and then led into canals beside the last-mentioned receptacle. Lastly, the substance is gathered into masses, from which are cut the parallelopipeds which, after dessication, are supplied to commerce. Caustic magnesia is obtained by heating the carbonate in red-hot muffled furnaces.

ADULTERATION OF WAX WITH TALLOW.

BY HARDY.—Wax floats in alcohol of 29 deg. By observing the strength of the alcohol in which the sample floats, the percentage of wax may be deduced.

If the alcoholometer shows	29	degs. the wax contains	100 per cent wax.
"	"	39.63 "	"	"	75	"	.
"	"	50.25 "	"	"	50	"
"	60.87 "	"	"	25	"
"	"	71.50 "	"	"	0	"

MANUFACTURE OF SULPHURIC ACID.

INSTEAD of the lumps of coke or balls of earthy material usually employed in the towers in which sulphuric acid is condensed, Joseph Saunders, of Brooklyn, N. Y., proposes to use hollow balls of glass about six inches in diameter, arranged with their openings on the top in the tower, in which the gases descend, and the reverse in the tower through which the gases ascend. As these glass balls are not affected by the acid, the annoyances resulting from the friability of the materials previously employed are obviated.

TESTING ALCOHOL.

It is customary to obtain the percentage of absolute alcohol and water in mixtures of alcohol by taking the specific gravity with a hydrometer especially adapted to the purpose and called an alcoholometer. When a liquor contains syrups and extractive matters, the specific gravity fails to indicate the amount of alcohol present. In such cases, it has been necessary to distill off the alcohol and then measure it.

In these cases, and also where no alcoholometer is at hand, or the quantity of the liquid is too small to float one. Vogel's method may be employed. He found that, when dry starch-paper was dipped into a solution of iodine in alcohol of 66.8 per cent or over, the starch was not turned blue. If the spirits contained less than 66.8 per cent absolute alcohol, the paper is immediately blued. To apply the test to weaker alcohols, it is only necessary to add absolute alcohol until the reaction no longer takes place. From the quantity added it is easy to calculate the percentage. If the spirit tested is above 66.8, water is added from a graduated measure until the starch-paper turns blue, and the percentage calculated from the quantity of water added. If potassium be thrown upon alcohol of specific gravity 0.830, it takes fire; but with spirits of specific gravity 0.823 and under, it will not take fire.

THE BITTER APPLE AS AN ARTICLE OF FOOD.

By F. A. Flueckiger.—The bitter apple, bitter cucumber, bitter gourd, or bitter colocynth (*Citrullus Colcynthis* Schrader) is a creeping cucurbitaceous plant which grows abundantly in the Sahara in Arabia, and on the Coromandel Coast, and is found in some of the islands of the Ægean Sea. The fruit, which is about as large as an orange, contains an extremely bitter and drastic pulp, from which colcynth is obtained. This pulp is said to be eaten by buffaloes and ostriches, but is quite unfit for human food. The seed-kernels, however, which contain but a very small quantity of bitter principle, are used as food by some of the natives of the African desert. For this purpose, the seeds are first freed from pulp by roasting and boiling, and subsequent treading in sacks, and then deprived of their coatings—which are also decidedly bitter—by grinding and winnowing.

MANUFACTURE OF CHLORATE OF POTASH.

To manufacture chlorate of potash on a large scale, it has been recommended by W. Hunt to adopt the following method: Milk of lime is made to trickle down over bricks placed in a tower, where it comes in contact with a continuous current of chlorine gas. Chlorate of lime is the chief product; and, by treating this with chloride of potassium, chlorate of potash is formed, which can be purified by crystallization.

NEW METHOD OF PREPARING CAUSTIC SODA.

THE crude lye is evaporated in cast iron boilers. At a certain heat, the cyanides contained in the pasty mass are decomposed, with escape of ammonia and decomposition of carbon. When this point is reached, the heat is raised to redness, and the mass becomes more fluid. A sheet iron cover is then fitted upon the boiler, provided with an opening through which enters an iron pipe. This is plunged into the mass, and air is forced in. The graphite which separates rises to the surface and may be collected. The mass is tested from time to time to see if the sulphur is perfectly oxydized. When this is the case, the blast is stopped. the mass allowed to become clear. and run off as usual.—*M. Helbig.*

MILK OF MAGNESIA.

EVERY physician, and almost everybody else, knows the value of magnesia as a medicine, especially in cases of "sour stomach," or, in more professional parlance, "indigestion"; but very few, even of the faculty, know how to administer it without creating more trouble than they relieve. The case has been that the only way of administering magnesia was in an almost insoluble powder, so that either children or adults were liable to be troubled with accumulations in the intestines of masses of the dried powder. Hence, notwithstanding its value as a medicine, magnesia has been ignored by physicians because they could find no certain method of giving it without trouble. The problem has at last been solved, however, by Mr. Chas. H. Phillips, who has perfected what he calls a "Milk of Magnesia," and what others may describe as a hydrate, perfectly soluble in water, and adapted to all the medical uses for which magnesia in any form is wanted. Many physicians have already tried it, and all coincide in endorsing the opinion we have expressed as to the merits of this preparation.

VALUE OF IMPORTED DRUGS, CHEMICALS, ETC.

The following will show the value of some of the larger articles of chemicals, &c., imported during 1873:

Article	Value
Acids	$379,647
Acetate of lime	20,254
Alkali	40,915
Albumen	135,417
…inu m	2,554
…os	21,140
…m	5,200
…inous cake	1?,554
…a, carb	79,196
…a, mur	11,487
…a, sal	52,207
…a, sulph	32,483
Annatto	28,278
Aniline	5,116
Aniline, oil of	2,927
Aniline …es	490,570
…nse seed	18,022
…dry	15,009
…nic	96,286
Arrowroot	1,768
Arnica flowers	34,880
…gs	1,578
…	1,090,896
Assafœtida	25,119
Asphaltum	32,676
…utine	21,598
Bark, calisaya	124,435
Bark, cascarilla	5,?34
Bark, Peruvian	?0,989
Barytes	5,1?0
Barytes, carbof	2,650
Barytes, su'p	26,071
Black salts	573
Bleaching powders	693,370
Borax	1,875
Brimstone	902,049
Butter of cocoa	1,545
Burgundy pitch	1,503
B…dn …ves	1,791
…inel	3,776
…ptior	63,040
…ffdies	7,?99
…Kins	19,181
Carmine	6,790
Chamomile flowers	8,842
Chlorodyne	1,397
…alk	27,903
Chicory	77,430
Chlor. hydrate	11,883
Cochineal	727,047
Cobalt	1,048
C…tqth	453
C…rdvo root	784
…rm …tar	400,131
Creosote	2?7
Cubebs	3,680
…Mir	39,133
…go	100
Cinh…	107,?35
…fish bone	3,093
Divi divi	17,906
Dragons' blood	1,322
Ergot of rye	3,568
Epsom salts	576
Extract of indigo	44,333
Extract …hdr	1?4,367
Extract o. …ker	15,387
Extract …nac	15,326
Fullers' earth	2,277
…ier	220,349
Gamboge	7,208
Garancine	1,350,790
Gentian root	4,554
Glue	9,?33
Glucose	2,?00
Gums, crude	46,662
Gum arabic	341,676
Gum aurine	2,946
Gum benjamin	2,636
Gum …ajvi	34,039
Gum …cpal	119,324
Gum damar	3,177
Gum gedda	3,722
Gum …gum	1,356
Gum kowrie	335,861
Gum olibanum	350
Gum senegal	5,620
Gum …harac	3,561
Gum …ha-nute	2,067
Gum talc	28,459
Gum tragacanth	29,215

Value of Imported Drugs, Chemicals, Etc.—Continued.

Article	Value	Article	Value	Article	Value
...lic.	...lic.	Oil cassia	...341	Oil ...ed	4,927
Gum dm	653	Oil ...ay	3,701	Oil ...nie	11,219
...fine	4.58	Oil cajeput	869	Oil ...nic	288
Indigo	503,709	Oil ...nn	373	Oils ...ind	8,580
Iodine	139,074	Oil citronella	34	Orchilla ...nl	906,142
Iode pot	630	Oil citron	853	...bella liqor	17,803
Iperac	19,118	Oil ...obs	...83		9,757
...let ...jar	8,297	Oil ...mnt	44,938	Orange peel	790
...lp	5,664	Oil cod	7,357	rebs root	648
Juniper bries	3,351	Oil croton	1,163	...de of zinc	198,456
...de dye	7,903	Oil fish	1,129	...ints	72,469
Laurel ...des	...726	Oil fusil	23	Paris ...hte	18,733
...hs	6,203	Oil ...pgli	478	Persian berries	684
Lavender ...hrs	048	Oil ...rm	534	...Ris	45,111
...lne ...ne	8,537	Oil ...ahm	3,140	...rtjo	395,047
...lne ...et	12	Oil ...jnr	1,507	...sh, bich	169,393
Locust ...bns	1,770	Oil ...nid	4,069	...tlh, hdor	1434
...Mr	91,610	Oil ...ahr	31,878	...th, mur	87789
...Ma	8,910	O laurel ...fon	394	...tlh, pes	104,397
Magnesia	2206	Oil myrbane	224,911	...tth, tnh	933
...Mg salt	28	Oil ...rge	7,198	...te, ...nph	104,643
...Mt	2,835	Oil olive	1336	...hr	135,949
Nitrate of ...hole	682	Oil opium	247,898	Rhubarb	61,331
Nitrate of lead	27,430	Oil palm	57,312	Saffron	23,608
Nutgalls	36,026	Oil ...rh	4,396	...Shr	12,181
Nux ...ma	3,425	Oil rose	1,129	Saltpetre	13,436
Oils ...nfol	78,751	Oil ...of	26,678	Santonine	383,338
Oil ...rse	22,749	Oil ...sue	995	...age 1 ...nss	2,82
Oil ...tie	1,373	Oil seal	4,064	Sarsaparilla	1,619
Oil ...hr	180	Ol sperm ...lad	16,042	Senna	90,215
Oil almond	8,332		95,025	Shellac	14,130
Oil ...rhot	191,560		5,407		275,136

Smalts	949	Sponges	121,881	Varnish	62,057
Soda ash	2,206,593	Squills	793	Vanilla beans	230,609
Soda, arsenate of	3,283	Sugar of milk	2,446	Venice turpentine	4,370
Soda, bi-carb	371,753	Storax	423	Verdigris	14,770
Soda, caustic	856,794	Sumac	275,525	Vermillion	64,028
Soda, hypo. sulph	15,346	Tonqua beans	42,231	Wormseed	6,246
Soda, nitrate	751,758	Turmeric	17,883	Yellow ochre	13,516
Soda, sal	366,079	Ultramarine	149,790	Yellow berries	58,734
Soda, silicate of	1,317	Valerian root	1,224	Other	263,853

CUSTOMS DECISIONS.

The following are the decisions and rulings of the Hon. Secretary of the Treasury during 1873 on all matters relating to Drugs, Chemicals, Oils, Paints, etc.:

IVORY DROP BLACK—DUTY ON ; TUSCAN RED, FREE.

CERTAIN so-called "ivory or bone black," commercially known as "ivory drop black," which was claimed to be free of duty under the provision of section 5, act June 6, 1872, for "bones burned, calcined, ground, or steamed," was properly classified for duty at the rate of 25 per cent. ad valorem, under the special provision therefor in section 10, act June 30, 1864.

Certain "Tuscan red" held to be exempt from duty, under the provision of section 5, act June 6, 1872, for "colcothar dry, or oxide of iron." —Letter to Collector at New York, January, 1873.

CARBONATE OF BARYTES—DUTY ON.

Carbonate of barytes, in a crude state, held not to be comprehended under the provision for "barytes," or "sulphate of barytes," in section 5, act July 14, 1862, it being the barytes earth, a raw and crude material from which the barytes of commerce is manufactured.

The article in question should be classified as a mineral substance, in a crude state, not otherwise provided for, at a duty of 20 per cent ad valorem, under the provision therefor in section 20, act March 2, 1861.—Letter to Collector at New York, January 1873.

ARROWROOT—DUTY ON.

The arrowroot of commerce, being in the form of a finely-pulverized starch, cannot be classified as exempt from duty under the provisions of section 5, act of June 6, 1872, for "root flour." It still continues to be dutiable under the special provision therefor in act of June 30, 1864.—Letter to Collector at New York, January 1873.

JAPANESE WAX—DUTY ON.

Japanese wax, being essentially different from the several kinds of wax specially enumerated as exempt from duty in section 5, act of June 6, 1872, cannot be classfied under such provision. The similitude clause of the act of 1842 does not operate to place an article otherwise dutiable on the free list.

By virtue of said clause, however, it was held to be dutiable under the provisions for beeswax, at 20 per cent. ad valorem.—Letter to Collector at New York, January 1873.

SINEWS—DUTY ON.

Sinews cannot be classified under the provision for "hide-cuttings" in section 5, act June 6, 1872. They are dutiable as a non-enumerated unmanufactured article at 10 per cent. ad valorem, under section 24, act March 2, 1861.—Letter to Collector at New York, January, 1873.

SIGNIFICATION OF TERM "BLACK SALTS."

The enumeration "black salts" in section 5, act of June 6, 1872, held to be limited to crude potash.—Letter to collector at New York January, 1873.

CERTAIN PREPARED CLAY—DUTY ON.

Certain so-called "prepared clay," bearing a close resemblance to French chalk, and used for the same purposes as that article, and being manufactured into small pieces about an inch in diameter, of different colors. was held to be dutiable at 20 per cent. ad valorem, under the provision for French chalk in section 9, act of June 30, 1864. or as an unenumerated manufactured article under section 24, act of March 2, 1861, and not, as claimed, under the provision for "unwrought clay, pipe clay, fire clay." &c., in said section, at $5 per ton, less 10 per cent. ad valorem, under section 2, act of June 6, 1872.—Letter to Collector at New York, January, 1873.

CASTILE SOAP—DUTY ON.

The Department holds that the provision of section 13, act June 30, 1864, for "all soap not otherwise provided for," does not operate to repeal the special provisions for castile soap in sections 22, act March 2, 1861, and 7, act July 14, 1862. Neither is castile soap embraced in the kinds of soap specially enumerated in the act of 1864 as dutiable at ten cents per pound and 25 per cent. ad valorem.

The article is consequently dutiable under the provisions above cited in the acts of 1861 and 1862 at 35 per cent. ad valorem.—Letter to Collector at New York, February, 1873.

DUTY ON CERTAIN SO-CALLED "BROWN GREASE."

Certain so-called "brown grease," being an oily compound or grease obtained by pressure from wool-skins, and very similar to sod-oil used by tanners and curriers for stuffing the pores of skins in the manufacture of leather, held to be dutiable either as a non-enumerated manufactured article under section 24, act March 2, 1861, or as an animal oil under section 21, act July 14, 1870, at the rate of 20 per cent. ad valorem.

The article in question not being such as is ordinarily fit for soap-stock, cannot properly be classified (as claimed) under the provision of section 5, act June 6, 1872, for "grease for use as soap-stock only, not otherwise provided for."—Letter to Collector at New York, February, 1873.

WHITE LEAD NOT ENTITLED TO TEN PER CENT. REDUCTION OF DUTY.

White lead is not entitled to 10 per cent. reduction of duty under the provision of section 2, act June 6, 1872, for "manufactures of metals, or of which either of them is the component part of chief value," it being held by the Department that such provision applies only where the metal entering into the composition of an article retains the physical form and properties of a metal.—Letter to Collector at New York, February, 1873.

The original precedent for the above decision is dated August 30, 1872.

DUTCH METAL AND BRONZE POWDER—DUTY ON.

In conformity with the recent decision of the United States circuit court for the southern district of New York, in the case of the United States vs. Ulman, the Department decides that hereafter duty will be collected on Dutch metal and bronze powder at the special rates provided therefor in the act, March 2, 1861, less 10 per cent., under act of June 6, 1872.—Letter to Collector at New York, dated March, 1873.

CHINA CLAY ENTITLED TO TEN PER CENT. REDUCTION OF DUTY.

China clay is entitled to 10 per cent. reduction of duty, under the provision of section 2, act June 6, 1872, for "*fine* clay."—Letter to Collector at Philadelphia, dated March, 1873.

FREE ENTRY OF BARRELS EXPORTED WITH PETROLEUM.

In pursuance of section 2, act March 3, 1872, authorizing the free entry of barrels of American manufacture, exported filled with domestic petroleum and returned empty, the following regulation is prescribed to carry said section into effect:

A declaration under oath, of the shipper at the foreign port, made before the United States consul at or nearest to such port, who shall certify thereto, if satisfied of the truth of such declaration, and authenticate the same by his consular seal, will be required, setting forth the fact that, to the best of his knowledge and belief, certain empty petroleum barrels [giving their number and description as far as practicable, or, if a copy of the invoice is annexed, stating, "described in the annexed invoice,"] are of American manufacture, and were exported from the United States filled with petroleum produced in the United States, and that it is intended to reship said barrels to the port of ——, in the United States, on board the [here the name of the vessel] now lying in the port of ——.

If such declaration is received by the collector of customs at the port of entry in the United States at the time the importation is made, and the party desiring to make entry shall, in addition, file his own affidavit with such collector, setting forth that to the best of his knowledge and belief the barrels of which it is desired to make free entry [describing them as far as practicable] and which were imported on the ----- day of ___ per ____, from ____, are of domestic origin, the collector at such port of entry in the United States, shall, if satisfied by examination of such barrels, that they are of domestic origin, admit the same to free entry.

If, however, such declaration of the foreign shipper is not received as aforesaid, and it shall appear from the affidavit of the party desiring to make entry, accompanied by such other evidence as the collector may require, that it is temporarily impracticable to produce such declaration, and that the same can be produced if time be given for that purpose, a bond may be taken from such party, with sufficient sureties, for six months, and in a penalty not less than the appraised value of the importation, for the production of the sworn declaration above required.

Upon execution of such bond and a compliance with the other portion of the regulations, free entry will be allowed.

These regulations will take effect on the 15th of June next, until which time customs officers will exercise such discretion in admitting such barrels to free entry as they may deem sufficient to protect the revenue.—Circular to Collectors, April, 1873.

COD SOUNDS—DUTY ON.

Cod sounds, in barrels, which it was claimed should be exempt from duty on the ground that they were imported for the purpose of being used in the manufacture of isinglass, were held to be dutiable at the rate of one and-a-half dollars per barrel, under the provision of section 10, act March 2, 1861, for "all other fish, pickled and in barrels," by virtue of the similitude clause of the act of August 30, 1842.

The claim of the appellants in this case could not be sustained, it appearing that sounds like those in question, which were preserved in salt and packed in barrels, are generally used as an article of food, and in such salted condition cannot be used in the manufacture of isinglass without great loss.—Letter to Collector, Boston, April, 1873.

PRUSSIAN BLUE—DUTY ON.

Prussian blue, which was claimed by the importers to be dutiable at the rate of 25 per cent. ad valorem, under provision of section 5, act July 14, 1862, for "paints and painter's colors, * * * dry or ground in oil," less 10 per cent., under the provision of the act June 6, 1872, for manufactures of metals, was held to be dutiable at 30 per cent. ad valorem, under the special provision therefor in section 11, act June 30, 1864, without the 10 per cent. reduction of duty.—Letter to Collector, New York, April, 1873.

LICORICE—DUTY ON.

The article in question was claimed by the importers to be "licorice juice," and, as such, primarily liable to duty under the special provision for licorice juice in section 5, act July 14, 1862.

It appearing, however, from the appraiser's return that the article is licorice in rolls, and that it is not in fact, nor commercially known as licorice juice, the Department held that it was dutiable under the special provision therefor in section 11, act June 30, 1864, at the rate of 10 cents per pound, less 10 per cent., under section 2, act June 6, 187', allowing that reduction on licorice paste.—Letter to Collector, New York, April, 1873.

PERMANGANATE OF POTASSA—DUTY ON.

Permanganate of poatssa, being a chemical salt, and used almost exclusively as a disinfectant, held to be dutiable at the rate of 20 per cent. ad valorem, under the provision of section 20, act March 2, 1861, for "all other salts and preparations of salts not otherwise provided for."—Letter to Collector, Philadelphia, May, 1873.

WHAT SHALL BE CLASSIFIED AS CREAM TARTAR UNDER THE TARIFF.

Any article which is in fact and substantially cream of tartar, and is used without further process of refinement for purposes for which cream of tartar is used, should be classified as cream of tartar, whether known by that name in commerce or by other designations, such as pink cream, &c.
—Letter to Collector, New York, May, 1873.

STANNATE OF SODA—DUTY ON.

Stannate of soda, being a compound of peroxide of tin and caustic soda, the chief value being the peroxide of tin, and like tin crystals and other salts of tin, of which the peroxide of tin is the component of chief value, being used as a mordant in fixing colors, was held to be dutiable, under the provision of section 8, act July 14, 1862, for "salts of tin," at the rate of 30 per cent ad valorem.—Letter to Collector, New York, June, 1873.

VENETIAN RED—DUTY ON.

Venetian red, although containing a percentage of oxide of iron, is not exempt from duty under the provision of section 5, act June 6, 1872, for "colcothar dry, or oxide of iron," it not being known commercially under any other designation than "venetian red." The article is consequently dutiable under the special provision for "venetian red," in section 5, act July 14, 1862, at the rate of 25 per cent. ad valorem.—Letter to Collector, Boston, June, 1873.

SEAL OIL NOT FREE OF DUTY UNDER THE TREATY OF WASHINGTON.

Seal oil imported from Canada is not exempt from duty under the provisions of the Treaty of Washington, for the free entry of fish oil, the article in question having been generally recognised in the tariff laws of the United States as a distinct article from fish oil, and not being in fact fish oil. Seal oil will, therefore, be subject to duty under the act of July 14, 1870.—Circular letter to Collectors, June 16, 1873. Customs division.

ENFLEURAGED OILS—DUTY ON.

Enfleuraged oils, which are highly perfumed oils of different orders, although not essential oils, and are perfumed by the same process as that by which pomades and hair oils are perfumed, and are used in the condition in which they are imported for the manufacture of perfumery, and for hair oil, were held to be dutiable under the provision for " hair oils, pomades, hair dressings, * * * or other perfumeries or cosmetics by whatsoever name or names known, used or applied as perfumes," in section 5, act July 14, 1862, at the rate of 50 per cent. ad valorem.—Letter to Collector, New York, June, 1873.

POWDERED PLUMBAGO EXEMPT FROM DUTY.

Powdered plumbago, although it had undergone a process of refinement for the purpose of removing any iron, lime, or other foreign substances that might have been present in the crude article as taken from the mines, was held to be "plumbago," within the meaning of the provision therefor, in section 5, act June, 1872, and therefore to be exempt from duty. A similar rule was laid down in the case of pumice-stone bricks.—Letter to Collector, Ogdensburg, July, 1873.

DUTCH-METAL SCRAPS—DUTY ON.

This article, being clippings or scraps of a manufactured article known as Dutch-metal, and not requiring manufacture, in the proper sense of that term, to render it fit for use, was held to be dutiable as a manufacture of brass not otherwise provided for, under the provisions therefor in the acts of 1861 and 1862, at the rate of 35 per cent. ad valorem, less 10 per cent. under the act of 1872.

For the reason above stated, the article cannot properly be classified for duty as old brass fit only to be manufactured.—Letter to Collector, New York, July, 1873.

POWDERED TALC EXEMPT FROM DUTY.

The article in question, though reduced to a powder, is commercially known and recognised as talc, and is therefore exempt from duty under the special provision for *talc* in section 5, act June 6, 1872.—Letter to Collector, New York, July, 1873.

OIL OF COAL TAR OR NITRO-BENZOLE—DUTY ON.

Oil of coal tar, also known as nitro-benzole, oil of mirbane, artificial essential oil of almonds, or essence of mirbane, and being the product of a mechanical mixture of benzole and nitric acid, held to be dutiable as a non-enumerated manufactured article, under section 24, act March 2, 1861, at the rate of 20 per cent. ad valorem.

This article, though not an essential oil, is used for the same purpose as essential oil of almonds, by reason of which fact it was classified by Department decision of April 22, 1868, with essential oil of almonds, by virtue of the similitude clause of the act of 1842.

Essential oil of almonds being now, however, on the free list (section 5, act June 6, 1872), such classification under the assimilating clause of the act of 1842 can no longer be adopted.

VICHY LOZENGES AND ASTHMA CIGARETTES—DUTY ON.

Vichy lozenges and asthma cigarettes, intended to be used medicinally, but not being proprietary, or prepared according to any private formula or secret art, held to be dutiable, under the provisions of the acts of 1861 and 1862 for "medicinal preparations not otherwise provided," at the rate of 40 per cent. ad valorem.

HAKE AND COD SOUNDS EXEMPT FROM DUTY.

Hake and cod sounds, being the air bladder of the fish dried, and in fact fish glue, requiring only to be softened and rolled to be fit for use, as isinglass, held to be exempt from duty under the provision of section 5 act June 6, 1872, "isinglass or fish glue."

This ruling does not conflict with that which relates to salted cod sounds.

FREE ENTRY OF WHALE AND SEAL OIL CAUGHT BY CER-TAIN AMERICAN VESSELS NEAR PUNTA BUNDA, MEXICO.

The Department regards whale and seal oil caught by American vessels of from ten to fifteen tons burden and licensed to engage in fisheries near Punta Bunda, Mexico, and near certain islands fifty to sixty miles off the coast of Mexico, as the product of American fisheries, and consequently free of duty, the parties engaged therein being American citizens and the vessels duly licensed. The fact that the vessels are under tonnage does not affect the right of the oil to free entry.—Letter to Collector, San Diego, Cal., August, 1873.

BRITISH COLUMBIA NOT ENTITLED TO BENEFITS OF THE WASHINGTON TREATY.

British Columbia not being a part of the Dominion of Canada at the time of the signing of the treaty between the United States and Great

Britain of May 8, 1871, is not entitled to the benefits of said treaty; and fish and fish oils, imported from that part of the Dominion of Canada, are not entitled to admission into the United States free of duty.

ANGOSTURA BITTERS---DUTY ON.

Angostura bitters held to be dutiable at the rate of $2 per proof gallon, under section 21, act of July 14, 1870, relating to "bitters containing spirits," every gauge or wine gallon to be counted, at least, as one proof gallon—this decision replacing the decisions of March 5, 1868, and November 16, 1869, that such bitters were liable to a duty of 100 per centum ad valorem and 50 cents per proof gallon.—Letter to Surveyor, Port of St. Louis, September, 1873.

BOTTLES CONTAINING COLOGNE WATER—DUTIABLE CHARACTER OF.

The Department decides that bottles containing Cologne water are not subject to a separate duty under the provision in section 17 of the act of March 2, 1861, for "all glass bottles or jars filled with sweetmeats, preserves, or other articles," which provision is held to apply solely to bottles containing sweetmeats, preserves, or other like articles; but that they are dutiable, as forming an element of the dutiable value of the Cologne water under the 9th section of the act of July 28, 1866, it being subject to an ad valorem duty.—Letter to Collector of Customs, Baltimore, October, 1873.

BREMEN BLUE—DUTY ON.

Bremen blue, so-called, claimed by the importers to be dutiable at the rate of 25 per centum ad valorem, less 10 per centum, as a "painter's" color, was held by the Department on the report of the appraiser that it was a mineral blue, to be liable to duty (under the special provision for that article in section 11 of the act of June 30, 1864) at the rate of 30 per centum ad valorem, and without the reduction prescribed in section 2 of the act of June 6, 1872.

SULPHATE OF AMMONIA—RATE OF DUTY ON.

Sulphate of ammonia, being specially enumerated and provided for in the 5th section of the act of July 14, 1862, at a duty of 20 per centum ad valorem, and being capable of use for other purposes than manure, is not entitled to free entry (although it may be intended for use as a fertilizer) under the provision in the 23d section of the act of March 2, 1861, for "substance expressly used for manure."—Letter to Collector of Customs, New York, October, 1873.

ALCOHOLADO—DUTY ON.

Alcoholado, so-called, which was ascertained to be not the bay rum of commerce, but a compound or preparation "of which distilled spirits are a component of chief value" (it being composed of alcohol, 92 degrees proof, and merely holding in solution a sufficient quantity of bay oil to give it an odor resembling bay rum), was held to be liable, by virtue of the 1st section of the act of July 28, 1866 (Heyl., 508), to a duty of $2 per gallon under the provision for "other spirits manufactured or distilled from grain or other materials and not herein otherwise provided for," contained in the 21st section of the act of July 14, 1870.—Letter to Collector of Customs, Providence, November, 1873.

DUTCH-METAL CLIPPINGS—DUTY ON.

Dutch-metal clippings made from copper held to be dutiable under the act of February 24, 1869, as a manufacture of copper, at the rate of 45 per cent. ad valorem, less 10 per cent., under section 2, act June 6, 1872.

This ruling does not affect Department's former decision of July, 1873, on Dutch-metal scraps or clippings, the article in question in that case, though commercially known as "Dutch metal clippings," being made from brass, and not from copper.

GLUE NOT ENTITLED TO REDUCTION OF DUTY UNDER SECTION 2, ACT JUNE 6, 1872.

Glue is not entitled to 10 per cent. reduction of duty under the provision of section 2, act June 6, 1872, for "manufactures of skins, bone, [or] horn," although in point of fact manufactured from those articles, the distinct character of such materials, when entering into the composition of glue, being lost by their assuming an entirely different appearance and shape.

The provision of law referred to applies only where the articles manufactured from "skins, bone, horn," &c., retain the physical properties and appearance of such component materials to such extent that the latter can readily be recognized from an inspection of the manufactured article. —Letter to Collector, New York, December, 1873.

DECISIONS OF THE

ARBITRATION COMMITTEE

OF THE

PRODUCE EXCHANGE

RELATING TO

OILS, NAVAL STORES, TALLOW, LARD, ETC., FOR 1873.

THE charter of the New York Produce Exchange was passed by the Legislature April 19, 1862, under the title of an Act to incorporate the New York Commercial Association, which act was amended February 13, 1868, changing the name of the New York Commercial Association to the New York Produce Exchange.

Sections 5, 6, 7 and 8 of the charter provides.

SEC. 5.—The Board of Managers shall annually elect, by ballot, five members of the Association, who shall not be members of the Board, as a committee to be known and styled the Arbitration Committee of the New York Commercial Association. The Board of Managers may, at any time, fill any vacancy or vacancies that may occur in said committee for the remainder of the term in which the same shall happen. It shall be the duty of said Arbitration Committee to hear and decide any controversy which may arise between the members of the said association, or any person claiming by, through, or under them, and as may be voluntarily submitted to said committee for arbitration; and such members and persons may, by an instrument in writing, signed by them and attested by a subscribing witness, agree to submit to the decision of such committee any such controversy which might be the subject of an action at law, or in equity, except claims of title to real estate or to any interest therein, and that a judgment of the Supreme Court shall be rendered upon the award made pursuant to such submission.

SEC. 6.—Such Arbitration Committee, or a majority of them, shall have power to appoint a time and place of hearing of any such controversy and adjourn the same from time to time as may be necessary. not beyond the day fixed in the submission for rendering their award, except by consent of parties; to issue subpœnas for the attendance of witnesses residing or being in the Metropolitan Police District. All the provisions contained in title 14, part 3d, chapter 8, of the Revised Statutes, and all acts amendatory or in substitution thereof, relating to issuing attachments to compel the attendance of witnesses, shall apply to proceedings had before the said Arbitration Committee. Witnesses so subpœnaed as aforesaid shall be entitled to the fees prescribed by law for witnesses in the Courts of Justices of the Peace.

SEC. 7.—Any number not less than a majority of all the members of the Arbitration Committee shall be competent to meet together and hear the proofs and allegations of the parties, and an award by a majority of those who shall have been present at the hearing of the proofs and allegations shall be deemed the award of the Arbitration Committee and shall be valid and binding on the parties thereto. Such award shall be made in writing, subscribed by the members of the committee concurring therein, and attested by a subscribing witness. Upon filing the submission and award in the office of the Clerk of the Supreme Court of the City and County of New York, both duly acknowledged or proved in the same manner as deeds are required to be acknowledged or proved in order to be recorded, a judgment may be entered therein according to the award, and shall be docketed, transcripts filed, and executions issued thereon, the same as authorized by law in regard to judgments in the Supreme Court. Judgments entered in conformity with such award shall not be subject to be removed, reversed, modified, or in any manner appealed from by the parties thereto, except for frauds, collusion, or corruption of said Arbitration Committee, or some member thereof.

SEC. 8.—This act shall take effect immediately.

On February 3, 1873, there was a case heard before the committee subject to the following agreement signed by the parties in dispute.

AN AGREEMENT

To submit to abide by the decision of the Arbitration Committee of the New York Produce Exchange.

We, the undersigned, members of the New York Produce Exchange, hereby agree to submit and do voluntarily submit to the Arbitration Committee of the New York Produce Exchange for their consideration and adjudication, a matter in dispute in reference to a contract for 2,000 bbls of Naphtha, dated November 6, 1872, made by R. W. Burke, and we hereby bind ourselves, heirs, executors or assigns, to abide by such decision as the said Committee may render; and in the event of a failure to comply with such decision as the said committee may render, we hereby authorize and empower the said committee to assess and award the damages arising therefrom, and we hereby further bind ourselves heirs, executors or assigns, to abide by such assessment and award as the said committee may render, and agree that a judgment of the Supreme Court be rendered upon the award made pursuant to this submission.

CHARLES PRATT.

SINCLAIRE & MARVIN.

Signed in presence of

HENRY H. ROGERS.

New York, January 31, 1873.

The case was stated by the plaintiff to be a failure in completing a contract made by R. W. Burke, to deliver 2,000 barrels Prime City Naphtha to Sinclaire & Marvin as per following contract put in evidence.

[FORM G.]

NEW YORK, November 6, 1872.

Sold for account of R. W. Burke, Esq., to Messrs. Sinclaire & Marvin, two thousand (ten per cent, more or less) barrels prime city naphtha, made from petroleum, (of the the production of the United States) color to be prime white and sweet, gravity sixty to seventy-three at eighteen (18) cents for each five and three-quarters of a pound, net, and three cents per barrel rolling charge. *Cash on delivery.* No charge for package, tare to be actual ascertainable as per condition ; no account hereupon endorsed. To be delivered in prime shipping order in yard, where sea-going vessels can load, or to vessel, seller paying lighterage subject to the conditions printed, on the back of this contract. Suitable to buyers vessel with ten long days, from 10th to 25th December, 1872. Barrels to be double glued. The three cents selling charge, to be refunded to shipper in case his vessel pays New York, or Brooklyn rates of wharfage. Brokerage one-half of one per cent by seller.

BABCOCK & COX, Brokers.

The plaintiff stated the goods were ready for delivery at the proper time, and no objection was made by the defendant on account of plaintiff assuming R. W. Burkes position in the delivery of the property, and they the plaintiffs asked defendants to send their vessel for goods as per contract, but the vessel was not sent, although defendant promised to do so, and finally on January 3, he refused to send vessel or receive the Naphtha, averring that the contract had expired, and that he was not obliged to receive it. The plaintiff further stated that it is customary in this branch of trade to exercise forbearance with each other in receiving and delivering goods, and he claimed that the indulgence shown toward defendant in this case should not be employed to deprive plaintiffs of their rights under this contract.

The defendants stated that the change of parties in the contract met no objection from them, but they would show that they were ready and wished to receive these goods, having different vessels loading this kind of goods from November to latter part of December. On 4th December asked Mr. Pratt to furnish the naphtha, which he declined to do, as the time was not yet up. He, the deponent, was therefore obliged to purchase to complete vessel, and paid $17\frac{1}{2}$ cents. On 24th December ordered a vessel to Pratt's yard, but it did not go, as the captain said that Pratt had two yards and he did not know which one to go to, so he went to neither. The matter received no further attention from him until the 5th of January, when plaintiff came to him about it, when he, the deponent, told him that the contract had expired, and was of no further force or effect. They having failed to make any tender of the property, it was too late now to do so.

The plaintiff, to show what is customary in the trade, presented Mr. Ackerman, of Messrs. Meissner, Ackerman & Co., as a witness, who stated that under a contract, buyer's option, it was customary for the buyer to give notice when he would be ready to receive on his contract, and that the clause in regard to an order on warehouse constituting a delivery was intended as a mode of settlement when the transaction was for speculation, and should construe the meaning of this contract to be that the vessel could be furnished by buyers on any day between the 10th and 25th of December, 1872, and that the seller had ten days to deliver next following the day said vessel reported ready for cargo at the place designated in contract.

Mr. Bunker stated that it is customary for buyer to give notice when his vessel is ready, but when parties are not known the seller would seek to effect a delivery and give notice, and it would be optional with a buyer to receive goods on a contract after it had expired, when no notice had been given by the seller. I should say this contract expired on 25th December. A seller would lose his rights in a contract if he neglected to make his claim at the expiration of a contract. The seller had ten days to deliver after the buyer reported his vessel ready for the goods. If the vessel reported on the 26th December, this contract would expire January 5th.

The plaintiffs further stated that they called upon the defendant on 23d December for a vessel to take the naphtha which he promised to furnish, but he did not. Saw defendant again on the 2d of January and asked him to send vessel, but he replied that he had cancelled contract. On the 5th of January he, plaintiff, took Mr. Burke with him and made demand on the defendant that he take delivery, which he refused, and they then mutually agreed to submit it to this committee for adjudication.

The plaintiff claimed that by the evidence it had been shown that the contract did not expire until January 5, and he had made demand on that day for the vessel to deliver the oil, which was refused, and we are still entitled to our rights under the contract.

The defendant claimed the abrogation of the contract on the legal point that there being a difference in the terms of the two contracts the law holds that they are void.

The evidence being all in, the committee present unanimously made the following award :

NEW YORK, Feb. 4, 1873.

Whereas a controversy between Mr. Charles Pratt and Messrs. Sinclaire & Marvin, members of the New York Produce Exchange, having been voluntarily submitted to the Arbitration Committee of the New York Produce Exchange for their decision by an instrument in writing bearing date the 31st of January, 1873, duly signed and attested, and whereas the proofs and allegations of the parties were heard at a meeting of a majority of all the members of said committee, and a majority of those present concurring in the following award :

The said Arbitration Committee of the New York Produce Exchange do hereby award, order and decide that Mr. Charles Pratt, having failed to tender the naphtha on the expiration of the contract, cannot now compel Messrs. Sinclaire & Marvin to receive the same, and he must pay the cost of this arbitration, twenty-three dollars ($23).

In witness whereof, we, members of the said Arbitration Committee, have hereunto subscribed our names.

(Signed.)   FREDERICK SHERWOOD,
             EDWARD HINCKEN,
             WALTER F. BRUSH,
             WM. BLANCHARD.

Signed in presence of JAMES BOUGHTON.

February 24, 1873.

A case was heard before the Arbitration Committee this day, being a matter in dispute in relation to sale of 215 tierces tallow, about 14th February, between Mr. G. M. Merrielees and Mr. D. Bauer.

The plaintiff stated that on or about the 14th of February he purchased from the defendant a lot of 215 tierces tallow, at 8 9-16c per pound, by a sample furnished by the defendant previous to the purchase, and when the defendant first mentioned this lot to him, it was with the idea of having him sell it for him as a broker. After trying for a while to do so without success, he proposed to take it himself. Some question arising in regard to the ownership, he, the defendant, said he would sell it himself to the plaintiff ; and he, the plaintiff, knew no other party in the transaction. Having the sample, he took it and sold it to Mr. Heydecker by the same sample, and Mr. Heydecker immediately engaged freight for the same.

The plaintiff went to the defendant to get the delivery order. The defendant said the lightermen had it aboard lighter and would deliver it free on board, or he would give him an order to deliver it.

On the following day being anxious to forward the delivery he called upon the defendant for an order, and he replied that he had given one order and did not wish to give another, but urging it, the defendant gave a written order, writing duplicate upon it, which was taken to the lighterman with vessel's permit attached, and was left with him to deliver the property on board. On going to see if the tallow had been delivered, was told by the lighterman that the defendant had countermanded the order and had taken it away. Thus he had been prevented delivering the tallow to the party to whom he, plaintiff, had sold it, whereby he had sustained loss for which he now claims as per bill herewith.

NEW YORK, February 14, 1873.

Messrs. D. BAUER & CO.

To G. M. MERRIELEES,   Dr.

To difference on 215 tierces tallow bought at 8 9-16c, sold at 8¾c. Estimated weight 6,900 lbs, at 3-16c, $129 39.

The plaintiff further stated that as there may other damages arise, not at this time known, he would give notice that further claims may be made if found to exist.

The defendant responded by saying that he had notice from Theo. Hermann & Co., New Orleans, of the consignment of 215 tierces tallow, which they wished him to sell for them, and saw Mr. Merrielees to have him do it for them. A sample was drawn promiscuously from the lot and given to Mr. Merrielees, and he knew at the time that the property did not belong to the defendant.

Subsequently, the plaintiff offered 8 9-16c, which defendant accepted and sent sale note to the owners, and they demurred at the responsibility of buyer. The plaintiff offered to give another name in the person of Mr. Burnett, and he then made mention of part of the lot having been rejected, to which defendant replied that the tallow was sold as a lot and could not be delivered in any other way, and he, the defendant, declined to deliver the tallow on these conditions, and went at once to the lighterman and recalled the order, as requested to do by the owners.

Cross examined: I gave the sample to plaintiff before the sale. The sale was based on the sample. The defendant did not know any more about the character of the tallow than plaintiff did.

The plaintiff further stated, that he bought the tallow for his own account, and not to apply on a contract, and sold it at a profit of 3-16c. per pound, which is the basis of this claim.

The defendant said he sold it on his own account.

The parties then retired, and the committee made the following award :

Whereas a controversy between Mr. G. M. Merrielees and Messrs. D. Bauer & Co., members of the New York Produce Exchange, having been voluntarily submitted to the Arbitration Committee of the New York Produce Exchange for their decision by an instrument in writing, bearing date the twentieth day of February, 1873, duly signed and attested, and whereas the proofs and allegations of the parties were heard at a meeting of a majority of all the members of said committee, and a majority of those present concurring in the following award : The said Arbitration Committee of the New York Produce Exchange do hereby award, order and decide that Messrs. D. Bauer & Co. must pay Mr. G. M. Merrielees one hundred and sixteen dollars and fifty-nine cents, for non-delivery of lot of tallow sold by them to Mr. Merrielees, and also pay the cost of this arbitration, $28.

Signed by members of the committee.

Meeting of Arbitration Committee held this day, at 3 P. M. All committee present.

A case submitted being a matter in dispute in regard to the settlement of a contract dated November 21, 1872, for the sale by Mr. James B. Grant of Cincinnati, to C. C. Abel & Co., of New York, of 250 tierces choice kettle rendered lard, new or old, with all the questions in controversy or dispute arising under or growing out of said contract.

The plaintiff stated that he bought through Mr. Morawitz, broker, 250 tierces lard from the defendant, of Cincinnati, as per following memorandum :

NEW YORK, November 6, 1873.

Messrs. C. C. ABEL & CO

Bought for your account from Mr. George Megrath, two hundred and fifty tierces (250) wooden bound choice Western lard, packing 1871-72, (new or old) at eight and three-fourths (8¾) cents per pound, and to be delivered free on board buyer's vessel, at seller's option, during the first ten days of next month (December), tares to be actual. Terms cash. Seller privilege of drawing sight draft with bill of lading, less freight and 5 per cent; brokerage by buyer.

S. MORAWITZ, Broker.

The plaintiff stated that early in December he received notice of arrival of several lots of this lard, which had been bought for shipment to Europe, and as the steamer was to sail on December 12th, he applied to railroad company to deliver their lard, as per bill of lading, to the steamer, furnishing them at the time steamer's permit, but the railroad company declined to deliver it, as the whole lot had not arrived, and would not deliver until they could deliver the whole consignment at once. On the 9th December lard had not all arrived, and, fearing the railroad would not deliver, the plaintiffs purchased 200 tierces lard for shipment by their steamer, and sent it aboard December 10, the day the contract expired. Having paid defendant's drafts for about $6,000, plaintiff drew repayment for said amount, accompanied by bill of lading enclosed, he, defendant, having failed to deliver, as per sale note, on his. plaintiff's, call on last day of contract. The bill of lading and draft was returned to them unpaid. They, the plaintiffs, took possession of defendant's lard and sold it for account of defendant, to reimburse themselves for their advance. They sold 200 tierces of it on December 16 (50 having been lost in transit), as follows: 149 to J. H. Pool, the remaining 51 plaintiffs assumed, allowing defendant 8¼c., the market being at 8c.

The defendant was not satisfied, and, Mr. Pool having the proceeds in his hands, it was agreed to submit it to arbitration, and the plaintiffs submit a claim of $608 87, as balance due them.

Mr. Pool stated on 23d January, 1873, received notice from railroad of the arrival of 50 tierces lard marked D, and the defendant telegraphed it was for him, and he sold it, and has proceeds.

Defendant stated he negotiated with plaintiff for settlement of said contract in case of failure to arrive in time by purchase of other lard, and have this sold at the market price. On the 9th, received telegraph from broker: Lard not arrived, bought 49 from Pool; you have to settle for. On the 10th telegraphed to broker: Let plaintiff replace all short and sell lard when it arrives, giving no limit. Received reply same day: Railroad will not deliver to-day; have replaced one fifty for one-quarter; will give 8¼c. for 100. Defendant telegraphed: Will sell 100 behind at quarter; offer for some more. The broker was acting for plaintiff. Defendant objected to charges for delivery. Plaintiff having taken position that the contract became void by reason of non-delivery on the 10th December. He, defendant, having provided for fulfillment of contract, and broker's telegram saying lard had been in to fill it. I claim it should be settled by charging defendant with all necessary cost for same, and crediting him with the contract price of lard.

Plaintiff further stated telegrams were unauthorized by him, and that the contract between them became void on 10th December, by reason of failure to deliver, and their charges were that of a consignee (which they assumed) for the care and sale of the property.

Broker testified that he was not authorized to telegraph by plaintiff, and did not understand at the time that the purchase was to apply on defendant's contract.

The committee, after hearing the above evidence, decided that the defendant must pay the plaintiffs five hundred and twenty-one dollars and thirty-seven cents in settlement of the account arising from the non-delivery by the defendant of two hundred and fifty tierces lard on his contract, dated November 6, 1872, and also pay costs of this arbitration, $25

Signed by committee.

NEW YORK, May 7, 1873.

Arbitration Committee meeting held this day, at 3 P. M., all the Committee present.

Case of David Dows & Co. *vs.* John F. Cook.

The plaintiffs stated that they sold to the defendants on April 16th, 250 tierces lard, at 9 cents, deliverable in April, which they held up to 23d April, when they delivered 83 tierces, to complete the lot; and the defendant has paid for the same in part, but refuses to settle the balance of $41 23, which is our claim.

The defendant said the lard had been delivered to Cragin & Co., who refused to settle on account of a difference on the tare claimed by Wilcox & Co., to whom they delivered it, and claimed that he was only a middleman, and would like to know if Mr. Wilcox occupies any different ground in this matter than I do, and if there be a difference, would like to know who is to pay it, and would like to have some decision made as to how such cases are to be adjusted.

The plaintiff replied that he made the contract in accordance with the rules of the Exchange, and would refer particularly to Rule 6 as regards Weights and Tares. I made the contract with the defendant, and have nothing to do with any one else. I notified the buyer of the weigher, as can be seen by the orders. I tendered the lard on the 22d, and heard no objection to the weighing until the 2d of May, and meanwhile the lard had been taken from the place of delivery and put in other hands.

The plaintiff said his claim was for a deduction of 3 3-20 pounds per tierce on 83 tierces lard.

The parties then retired, and the Committee made the following award: That Mr. John F. Cook pay to Messrs. David Dows & Co. forty-one dollars and twenty-three cents ($41.23), and also the cost of this arbitration, $25.

NEW YORK, May 20, 1873.

Arbitration Committee meeting held this day, at 3 P. M., all the Committee present.

Case of Cragin & Co. *vs.* George Megrath.

The plaintiffs stated that they sold, through a broker, to George Megrath, 500 tierces Western lard, at 9½ cents, to be delivered in May in store. On the 12th we delivered him 73 tierces and gave him a warehouse order for the balance of 250 tierces. On the 15th tendered him the other 250 tierces, but on the 16th the defendant refused to receive it or pay for the first 250 tierces.

The defendant stated that the broker sold him 500 tierces of lard, and the only question is as to whether it is standard lard or not, and if the plaintiffs will give me a certificate that it was packed before March 1st, I will take it. I bought the lard and gave 1-16 cent per pound more for it on account of its being a desirable brand, and did not think it necessary to examine it, but on the next day heard that there was no such brand in the market.

The plaintiff claimed that it was sold as a specific lot in store on the 12th of May, and he had ample time to examine it, but on the 16th he refused it, and I claim that it was not sold under contract, and do not feel called upon to furnish a certificate, but so far as I know it is winter lard; and that it must be standard, I claim has nothing to do with the sale.

The defendants here presented as witnesses Messrs. Kingar, Sinclair, Harrison, and Dally, who testified that under the sale note presented winter lard was called for, and that a person would not be justified in tendering anything but standard lard.

The Committee, after hearing all the evidence, made the following award: The first 250 tierces of lard have been accepted and delivered and must be paid for in full, and that Mr. Megrath must also receive the remaining 250 tierces of the 500 on being furnished by Cragin & Co. with evidence that it is winter rendered, their own certificate to that effect to be deemed sufficient, and that Mr. Megrath pay the expense of this arbitration ($25).

————

NEW YORK, June 20, 1873.

Arbitration Committee meeting held this day at 3 P.M. Committee all present. Case of J. T. Davies vs. Eric Railway Company.

The plaintiff presented Bill of Lading, dated January 25, 1873, for 500 tierces of lard marked F. & Bros., New York, to notify George Megrath, 115 Broad street, order of W. S. Bentley, and stated that he had only received 258 tierces of it, and was short 242, and the only reply we ever received from the Railway Company is that it has been delivered, and our parties West informed us that it was in New York, and we then traced it to Fowler Bros., who admitted that they were long of lard in this same quantity as we were short, but refused to adjust the difficulty, and last April when I had contracts to fill was forced to purchase my own lard from Messrs. Fowler Bros., at 9⅞c per lb., the highest point of the market, and Messrs. Fowler Bros. knew that we were receiving lard from the same shipper and of the same mark, viz., F. & Bros. The excuse given by the railway company for not delivering our lard was that all lard was blocked up at Indianapolis, and during that time they were delivering lard to Fowler Bros. Ten days previous to March 1st our bill of lading was sent to Erie Railway, and I claim they should have notified George Megrath, as per directions in bill of lading. They offered me diamond M lard. which I did not take, as they wished me to guarantee them, which I did not care to do, as it was some one else's property. My claim is as follows, being a loser to that extent:

THE ERIE AND PACIFIC DISPATCH CO.

To JOHN T. DAVIES, Dr.

To 242 tierces lard short delivered, on bill of lading dated St. Louis. January 25, shipped by W. S. Bentley, weighing 78,820 lbs., at 9⅞c., $7,783 48.

The defendants admitted that they were indebted to the plaintiff for 242 tierces of lard, but object to the claim as presented on account of price, and claim that they should pay plaintiff at price of lard at time of delivery, and if his bill of lading had been presented in time he could have got his lard. The way bill does not say who to notify, but simply to the order of W. S. Bentley. who is the shipper at the West, and we are receiving large quantities of lard from him with the same mark, all for Fowler Bros, and we had no way of knowing until we received plaintiff's bills of lading, which was after we had delivered to Fowler Brothers. The fault is occasioned by shippers in the West, who ship a lot of the same mark. which is intended for different parties, but making no mention of that fact, and it is impossible for us to know. We were receiving a great deal of lard at the time for Fowler Bros., which was all marked F. & Bro., and we delivered to them all that came of that mark, as we knew them to be responsible parties, and we were guaranteed from all loss. We admit our responsibility to Mr. Davis, but claim that Messrs. Fowler Bros. are responsible to us, but we do not think we are responsible for more than the invoice price of the lard as price at time of delivery in New York.

The delivery clerk, Mr. Antis, of the Erie Railway stated that the plaintiff did not show bill of lading until long after Fowler Bros. had received the lard, and there were no directions on the way bill to notify Geo. Megrath. As soon as mistake was known I did all I could to rectify it, and offered them diamond M lard, which they refused to take, and claimed Fletcher lard, which it was impossible for us to give them. Part of the lot claimed was way billed diamond M, and they could have got F. & Bros.' lard if they had presented bill of lading in time. At the end of the season we found 239 tierces had been over-delivered to Fowler Bros.

The parties then retired, and the Committee made the following award: "That the Erie Railway Company pay the bill of Mr. J. T. Davies, as rendered, viz.: Seven thousand seven hundred and eighty-three dollars and fifty-eight cents, with interest from date of purchase, April 24, 1873· and also cost of this arbitration, $25.

NEW YORK, June 23, 1873.

Arbitration Committee met this day at 3 P. M., all present.

The case of Erie Railway Company vs. Fowler Brothers.

The plaintiffs, represented by Mr. Davis, stated that they made the award given by the Committee on the 20th in the case of J. T. Davies vs. Erie Railway as the basis of their claim against Fowler Brothers as what we owe J. T. Davies the defendants in this case owes us, as they took lard that belonged to Davies, which they had no right to do, and with the understanding that they would make it right if any mistake occurred. They were notified by us of their mistake, and Fowler Brothers' man said he would deliver the same number of packages to Davies from the dock to make it right, which he did not do. Davies purchased from Fowler Bros. lard; if not the identical packages belonging to him, it was of the same brand, and Fowler Bros., by taking Davies' lard, were able to fill orders at old prices, which he could not have done had he not taken the lard which did not belong to him, and we think we have a valid claim for the amount presented, viz., seven thousand seven hundred and eighty-three dollars and forty-eight cents ($7,783.48). A bill of lading being issued in the West. the cars are branded and each one has a way-bill, and if a way-bill like the one shown comes in we send out notice to the party receiving this brand of goods, and do not always require bill of lading when we know the parties are responsible, like the defendants; and it is also in order to accommodate them, and to have the freight removed as soon as possible, and when all is delivered the bill of lading is surrendered. Messrs. Fowler Bros. knew how much they should have received, and they should now make it right.

Mr. Antis, Delivery Clerk, stated that they found out shortly after that he had delivered the defendants too much lard, and told their representative, and he said he would make them (J. T. Davies) right. We delivered to defendant, without Bill of Lading, as we had faith in them, and a promise from them to guarantee us from any loss if occasioned by their getting too much lard. They have presented different Bills of Lading at different times for same brand of lard, and we considered that all lard marked F. & B. was for them. About March 1st Davies presented Bill of Lading, which was after the defendants had the lard, and I tried all I could to rectify the mistake. The lard did not all come on one way bill, and 500 tierces were way-billed diamond M, all to the order of W. S. Bartley.

The plaintiffs further stated, through Messrs. Clark and Foley, that F. & Bros. brand was always considered to be Fowler Bros., and so anxious have defendants been to obtain lard that they would take it at times before notice was given, and before the freight was paid. We only claim 239 tierces lard, as we lost 3 on the way which we are willing to pay for.

The defendants replied that at that time they were receiving a great deal of lard of about six different marks, and of all that number we have not received one lot correct, being short on some and long on others, and not correct as to weight. We received notice to remove our goods which arrived in various lots in December, January and February, and it did not come as shipped, but two or three different brands being in a car, and we took it just as it came, and it was impossible for us to tell then whether we were short or long of lard.

Mr. Megrath asked me in May if I was long of lard, and our cashier informed me that we were long 233 tierces, this was in the middle of May, when we balanced our books, and I told Mr. Megrath that when we made up our stock list we would give him the proceeds and in June the Erie Railway Co., claimed to have traced 239 tierces to us, 3 having been lost on the way and we notified the Erie Railway Company that we were long in lard, and we offered to give them the price received on contract for the 233 tierces, or price of the lard on day on which the mistake occurred which they refused to accept. Up to March 1st we have received lard from Erie Railway that was shipped in December, and should have been here in 10 days. We are now short on diamond M lard, and have not been able to check a single invoice of our weights this season on account of the way in which the goods have been delivered, and would say that during the past season the Erie Railway Company have done their business in a most careless manner. We did not: nor give the authority to any one, to guarantee the Erie Railway Company against any loss that might occur.

The defendants clerk testified that he receipted for the lard, and also had bills of lading for it, and had whatever quantity I received endorsed on the back, in one case the Bill of Lading was delayed, but I presented them before I took the lard away. The fault was in the West as some came Fowler Bros. to notify Fowler Bros,, and others F. Bros. with no instructions who to notify, and as we did not know of any one else receiving lard of the same mark we took it, and the Railway Company made no objections. I did not guarantee anything to the Erie Railway, neither did I agree to give Mr. Davies lard to make him square.

In March I saw that we were over and notified our cashier, and in February I heard of Mr. Davies' bill of lading, but did not see it. When I went to the railroad to take charge of our lard the man who had charge previously had been killed, and everything was in confusion, and it was impossible for me to know anything about the previous business.

The Committee, after hearing the evidence presented, decided that Messrs. Fowler Bros. pay Erie Railway for two hundred and thirty-nine tierces laid (239), seven thousand six hundred and eighty-seven dollars, ($7,687); less expenses one hundred and nineteen dollars and fifty cents, ($119,50), making seven thousand five hundred and sixty-seven dollars and fifty cents, ($7.567.50), with interest from April 24th, 1873, and cost of this arbitration, $25.

NEW YORK, August 25, 1873.

Meeting of the Arbitration Committee held this day at 3 P. M. Present: Messrs. Sherwood, Blanchard, Jewell and Dally.

Case of C. Menelas vs. Butchers' Hide and Melting Association in regard to purchase of 100 hogsheads tallow.

The plaintiff stated that he bought from the defendants 100 hogsheads tallow as prime, through a broker, Mr. G. Vandenhove. The tallow was shipped to London, and was proven to be not prime tallow, but had a mixture of lard and pig's fat in it, as stated in letters received from there (put in evidence). We had a sample sent to the Pharmaceutical Society of Great Britain, and had it analyzed by Prof. Redwood, whose statement is submitted, testifying that part of the one hundred hogsheads contained a mixture of soft fat, such as lard. I claim $827 93 damages for inferior quality tallow sold me. I sent them a letter, dated July 9, requesting a settlement, to which they replied they would pay nothing. I have received letters this morning from the other side saying the tallow was not yet sold.

The defendants stated that they sold to Mr. Menelas, through broker, the tallow in question, which was tried by the broker, and it was all right. We do not make anything but prime tallow, and melt nothing but beef and mutton fat, right from the cattle, and never have any lard or pig's fat in our place. When we have scorched tallow we never sell it as prime tallow. Mr. Menelas made a claim, which was placed before our Board of Trustees, and by their request I informed Mr. Menelas, through Mr. Vandenhove, the broker.

The defendants' foreman stated substantially the same as the preceding witness.

The broker stated that he bought the tallow as prime tallow, sampled it and found it all right from his knowledge of tallow as a broker. If there was 5 to 6 per cent lard in it it could not be detected except by analysis, to which it is sometimes submitted at the buyer's expense, if he desires it.

The testimony being all in, the Committee made the following award:

That, according to the evidence submitted, Mr. C. Menelas has not established any claim against the Butchers' Hide and Melting Association, and must pay the cost of this arbitration, $20.

NEW YORK, October 15, 1873.

Meeting of the Arbitration Committee held this day at 3 P. M. Present: Messrs. Sherwood, Hincken and Jewell.

A matter in dispute for a non-fulfillment of contract for delivery of a cargo of rosin between Messrs. Julius Hess & Co., as plaintiffs, and Messrs Barker Bros., as defendants.

The plaintiffs stated that they bought on July 2, 1873, from the defendants (as per contract put in evidence) 1,600 barrels Wilmington strained rosin (10 per cent more or less), at $2 40 per 280 pounds, delivered free on board vessel "Three Sisters" upon her arrival at Wilmington, at the price of 7s. sterling freight per 310 lbs. The vessel arrived in due time, and all they could put upon her was 1,289 barrels, which, according to our contract' was 241 barrels short, and as we sold it upon the other side upon the same terms as we had bought it, and were obliged to cover ourselves in Hamburg for what we were short at a loss to us of $50 41, which we present as our claim, more as a matter of principle than for the amount of money involved. We obtained both the rosin and vessel from the defendants, and they represented she was to carry 1.600 barrels, 10 per cent, more or less. We have demanded fulfillment of contract from them, but they refuse to settle.

The defendant stated that the intention was to sell a cargo for the "Three Sisters," which we have delivered to them, and have put one hundred barrels on the deck in addition, and would have given her more if she would have held it. The words "as vessel may carry" are erased upon the contract, which was done without my knowledge, and must have been done by the broker. We chartered the vessel and sold both together, and claim by giving her a full cargo we have fulfilled the contract.

Mr. Halpin, the broker, stated that the defendant had the vessel "Three Sisters," and asked me to sell a cargo for her and named price of vessel and rosin. The vessel was 146 tons, and we estimated 10 barrels to a ton and thought we were safe, as we find small vessels carry more in proportion to their size than large ones. It was the intention to sell a cargo by the "Three Sisters," and all I can say is that there was a mistake made in estimating, and it is the fault of the one making it. It would not have been a sale if the vessel had not arrived, and if rosin had gone down to $2 the defendant would have had no right to tender the plaintiff the 141 barrels, and if the vessel had held 2,000 barrels the defendant would not have been obliged to give them that amount, but only 10 per cent. more than the 1,600 barrels. I do not know who altered it, and cannot say why it was altered.

The defendant further stated that he did not receive a copy of the contract from Hess & Co. until a week after it was signed. I have one contract for 2,000, on which I delivered 1,696 and nothing was said, and another contract for 1,100 on which I delivered 871, and the parties refused to take any more, and such things are happening every day. They claimed 141 barrels with 7s. freight, and my contract says nothing of the kind, and I do not think it had that erasure on it when signed by us.

The plaintiff stated that he had the clause (as vessel may carry) stricken out, and would not have received contract unless it had been stricken out, and I told them at the time I would claim the rosin if the vessel did not arrive, she being at sea at the time, which I hold under that contract I had a right to do. The reason of the contract not being signed was on account of the head of the house being away.

The defendant's clerk testified that he signed the contract and it was not altered then, and it was afterward passed by Mr. Barker, and it must have been altered between the time it was signed and the time it was passed.

The evidence being all in, the parties retired and the Committee made the following award: That, according to the evidence submitted, Messrs. Barker Brothers must pay the claim as presented by J. Hess & Co., for fifty dollars and forty-one cents ($50 41), and also pay the cost of this arbitration, $15.

In witness whereof, we, members of the said Arbitration Committee concurring therein, have hereunto subscribed our names this 15th day of October, 1873.

FREDERICK SHERWOOD,
EDWARD HINCKEN,
A. S. JEWELL.

Signed in presence of S. H. GRANT, Supt.

November 10, 1873.

Meeting of the Arbitration Committee held this day at 3 P. M. Present: Messrs. Sherwood, Hincken, Blanchard and Jewell. Mr. Sherwood in the chair.

A matter in dispute in regard to the purchase of 400 tierces of lard between Messrs. W. J. Wilcox & Co. and Mr. Albert C. Oertel.

The plaintiffs stated that on October 29th they sold to defendant 400 tierces lard, through Mr. F. Schnitzspahn, broker, as per contract put in evidence. They delivered the lard according to contract, and presented the bill on the next morning with the receipts and demanded payment, which was refused, and, sooner than sell him out, we offered to carry it for him until the following Wednesday (this was on Saturday) upon payment of $1.000 cash, free of interest, which he would not do, and we also offered to take his check, dated Monday, and in fact did everything we could to accommodate him. We offered to keep it open until Monday if Mr. Braun, to whom we sold it afterwards, would keep his offer open until that time; but Mr. Braun told us he could not keep it open, and we sold it to him on Monday noon at eight cents, which was the highest price paid for lard on that day, as we can prove by our sales to other parties. Mr. Braun was the only person who had orders for the Stettin steamer on that day, and we sold it to him, and it would have been a great sacrifice to us to have sold it in the open market. We claim a difference of one-quarter of a cent per pound on 400 tierces, $315 53. We always collect on delivery of receipts.

The defendant testified that it was a new business to him, and he was recommended to Mr. Schnitzspahn, the broker, and I gave him orders to purchase. He reported that he had purchased at a reasonable figure, and that, according to permit, plaintiffs would not ship it until Saturday, and payment would not be demanded until the following Wednesday. Plaintiffs shipped the lard on Saturday and sent the bills and receipts to me on the same day about 12 o'clock and demanded money, and I told them I was not prepared to pay, as it was so difficult to sell exchange, and I had carefully forwarded my exchange with Duncan Sherman & Co., and was prepared to pay on Wednesday, according to agreement as made with the broker. Mr. Van Ingen, clerk of the plaintiffs, came to my office on Monday, and asked me what was the matter in regard to the lard, and I told him I would pay for it on Wednesday, and he then said he would sell at once, a buyer being then at the office. I told him of the agreement made with the broker, and he said he did not care for any agreement made with him. I went and saw the plaintiffs, and they offered to carry it until Wednesday if I would put up a margin of $1,000, which I refused to do, as I was not obliged to put up any margin, and the plaintiffs further said that there was a man then waiting to buy the lard. They did not sell the lard on Monday at 12 o'clock, as on that day between 4 and 5 o'clock they sent to my office and asked me if I would pay on Wednesday, and I told him I would do what was right, and, if they had sold the lard on Monday, why did they send to me on the afternoon of that day. Tuesday, unfortunately for me, was a holiday, and I could not sell my exchange. When I was at plaintiffs' office they said they could transfer the lard without any cost to me, and afterwards said they had sold it at 8½c.

Mr. Van Ingen stated that the lard was shipped on Friday night, and the permit is dated October 31st. I demanded money on Saturday on delivery of receipts, and he refused to pay, and I told him we would exact payment before the steamer sailed or we would sell him out, and we offered to carry him until Wednesday upon payment of $1,000, and called upon him Saturday afternoon when he refused to pay, and I kept it open for him until Monday upon my own responsibility, and saw him on that day about 12 o'clock and asked him to accept the receipts, and he said he could not pay, and left me very abruptly. I also asked him if he would take care of it by Wednesday, but he would not promise, so I turned over the receipts and bill to Mr. Braun, and he paid me for the lard, and he was the only buyer that I knew of for lard for the Stettin steamer.

Mr. Braun testified that he bought the 400 tierces on Monday at 8c, and received my receipts about 4 P. M. I was offered this lard on Saturday at noon, but with the proviso that I would have to wait until Monday before it could be definitely agreed upon. I bought 200 tierces from plaintiffs on same day at 8c, which was the highest price paid.

The broker testified that he bought the lard and told the defendant that the money might be asked for on Saturday, and said nothing about his not having to pay until Wednesday, and am not authorized, and have never made any agreement outside of the contract.

The defendant further stated that the money was not due until Wednesday, and plaintiffs had a right to keep the receipts until the goods were paid for, and I told plaintiffs if the lard was delivered on Saturday I could not pay until Wednesday.

The parties then retired and the following award was made by the committee: That in accordance with the evidence presented Mr. Albert C. Oeitel must pay the claim of Messrs. W. J. Wilcox & Co., as presented, for three hundred and fifteen dollars and fifty-three cents ($315 53), and also pay the cost of this arbitration, $20.

In witness whereof, we, members of the said Arbitration Committee concurring therein, have hereunto subscribed our names this 11th day of November, 1873.

FREDERICK SHERWOOD,
EDWARD HINCKEN,
A. S. JEWELL,
WILLIAM BLANCHARD.

Signed in presence of WILLIAM E. FLETCHER.

NEW YORK, November 25th, 1873.

Meeting of the Arbitration Committee held this day at 3 P. M. Present: Messrs. Sherwood, Hincken, Blanchard, Jewell and Dally. Mr. Sherwood in the chair.

The chairman stated that this was a matter in dispute in regard to a difference with reference to the sale of petroleum tar or residuum between Messrs. Sone & Fleming and Messrs. Libby & Clark.

The defendants stated that they were not ready to proceed with the case, as they required considerable testimony, there being a large amount of damage involved. We have foreign testimony to introduce which is necessary to establish our case, and do not wish to be understood by the committee that we ask for unreasonable time, and will be prepared to go on with the case after the 10th of December. The case is purely one of damages, and extends over a period of twelve months, and as Messrs. Sone & Fleming will claim a large amount, we think that we are only asking for reasonable time.

The plaintiffs stated that they were prepared to go on, and would state their case in a few moments as follows : On the 8th March, 1873, we made two contracts with the defendants, they to take all the residuum made at our works (as per contracts put in evidence), one from March 8th, 1873, to October 1st, 1873, and one from October 1st, 1873, to March 8th, 1874. They took the tar for about two months, in all 4,280 barrels, and then rejected it, and we have since run all the tar at our works for their account at a loss to us, and estimate our damages to be about $30,000 to date, and as one of the contracts has not yet expired, we cannot tell what our total damages may be. We do not think foreign evidence has any bearing upon the case, and we claim fulfilment of the contract and refer to the rules of the petroleum trade, and would also call the attention of the committee to the inspector's certificate attached to the contracts, among them certificates of their own inspector, and we also have ten to fifteen more. I will present the affidavits of twenty-six of our men testifying that we have made residuum during the past year precisely as we have made it for the past twenty-two months. We have delivered this residuum as a merchantable article to Meissner, Ackermann & Co., and other large shippers. We have sold defendants tar prior to this contract. The defendants replied that they had everything to prove, and required time to do it. We claim that we have not accepted any of this tar except under protest, and also propose to show that it is not tar. As to the rules of the petroleum trade to which the plaintiff refers, I will state that they were not in force at that time and have been adopted since the contract was made, and the plaintiff was one of the committee to make said rules. We have a very valid answer to all the plaintiff's statements. To shorten and simplify the case we will present the following paper (read) which, if the plaintiffs will sign, admitting certain points, it will go far towards shortening the case. If we can demonstrate according to equity that we are entitled to reject the residuum, then we are entitled to damages, and if we cannot the plaintiffs are entitled to damages.

By plaintiffs: We refuse to sign the paper read by the defendant. They claim that residuum is only used for one purpose—that of making lubricating oil. We claim it is not so, as there were a half million barrels of illuminating oil made from it last year, and it is also used in making aniline colors, gas. and for other purposes, it being but a small portion that is used for making lubricating oil; and, besides, we claim it is not our business to know what it is to be used for. There have been cargoes of it shipped to Liverpool lately, and we have never sold it to refineries until the market for crude oil became so low.

By the defendant: The article the plaintiffs talk about our buying prior to this contract was bought for experiment, and sold to us under the name of 10 per cent tar, and we tried it and rejected it, and had no idea that such an article was to be tendered under this contract. We did not use a barrel of the tar, and we propose to prove what this article is by those who understand the qualities of residuum, and will have very elaborate proofs. There has been but very little shipped until this year, and it has but one use there as it has here. The United States Government decision says that under 20 gravity is tar, and over 20 burning oil; and, as a matter of fact, we shall demonstrate that other residuum is worth more than theirs, and that theirs is entirely unsuitable for our business. We have a great many witnesses, some of whom are out of town, and would ask to be allowed until the 15th December to obtain our evidence.

By the plaintiffs: I will admit that a heavy gravity tar may be an advantage in making lubricating oil; but we do not specify gravity, and sell the residuum as it is, having never put a hydrometer in it. We make one-third of all that is made in New York and vicinity, and claim that it is worth 50 per cent more than any other. The market has gone down; but if it had gone up, and I should have gone to the defendants and told them that I would not deliver them any more tar, as I had found out it was not suitable for making the article that they had bought it for, I think they would have said we can ship it, and wish it delivered according to contract. We do not object to reasonable time being given the defendants, but we are running from 1,200 to 1,500 barrels of tar a week, which makes a great loss to us.

The Committee, after consultation, stated that they were willing to give the parties all reasonable time, and would adjourn the case until Monday, December 15, 1873, at 3 P. M., Messrs. Libby & Clark to pay the cost of this adjournment, $25, and they hoped the parties would come prepared to proceed with the case as rapidly as possible.

On motion, adjourned.

WM. E. FLETCHER, Secretary.

NEW YORK. December 15th, 1873.

Adjourned meeting of the Arbitration Committee held this day at 3 P. M. All the committee present. Mr. Sherwood in the chair.

The chairman stated that this was an adjourned meeting of a matter in dispute between Messrs. Sone & Fleming and Messrs. Libby & Clark, in regard to a difference with reference to the sale of petroleum tar or residuum.

The plaintiffs submitted two contracts as follows: one from March 8th, 1873, to October 1st, 1873, and one from October 1st, 1873, to March 8th, 1874. they both being similar with the exception of a difference of $\frac{1}{4}$c per gallon in price.

CONTRACT.

NEW YORK, March 8th, 1873.

Sold to Messrs. Libby & Clark for account of Messrs. Sone & Fleming, all of the petroleum tar produced at their refinery, "Kings County Oil Works," from date to October 1st, 1873, at (6¾) six and three-quarter cents per gallon in bulk, by guage. Tar to be free from water and coke. Subject to inspection. Terms cash. To be delivered to lighter at sellers' yard in lots of not less than 200 barrels each. Sellers to furnish merchantable order barrels for transportation, and buyers to take away the tar within five days' notice from sellers, unless prevented by the elements or inability to procure lighters. Sellers reserving right to supply lighters at customary rates; buyers to return barrels of equal quality promptly. as required by seller, free of lighterage.

TRUMBY & BECKWITH,
Brokers, 134 Maiden Lane.

Brokerage by seller. Accepted.

LIBBY & CLARK.

March 11th, 1873.

Plaintiffs presented their claim for damage sustained on first contract, marked Exhibit 1, for $23,666 46.

The defendants submitted their claim for damages amounting to $40,000, marked Exhibit No. 2, and stated that they were prepared to prove that such damage was caused them by the neglect of the plaintiffs in not furnishing us a merchantable tar. Petroleum tar is the heavy residuum left in the stills from the distillation of crude oil, and should be from $2\frac{1}{2}$ to 5 per cent. crude oil, and from 16 to 20 specific gravity, while the tar that the plaintiffs have tendered us is unlike petroleum tar, containing 10 per cent. crude oil, and being from 22 to 26 gravity, and it should be properly classed as 10 per cent. tar, and which tar is totally unfit for our business of making paraffine oil and wax. This tar was bought subject to our inspection, and all that we have accepted has been under protest, and have refused to receive any more at all, and it is for the committee to decide as to whether those rejections were honest or not. We are prepared to prove by the unanimous testimony of the trade that petroleum tar should not be more than 2½ to 5 per cent. crude oil, and from 16 to 20 degrees specific gravity, and as we are properly the plaintiffs in this matter, and have everything to prove, we will, with the consent of the committee and of the plaintiffs, now call Mr. H. H. Rogers, of the firm of Charles Pratt, to the stand as a witness.

Mr. H. H. Rogers's statement: I have been a refiner of petroleum for years. (Contract shown, and question asked witness as to what kind of tar would be a fair tender on it.) Answer: I should say that contract means the customary tar which is from 3 to 5 per cent. crude oil, not over 5 per cent. and under 20 degrees specific gravity, and I should always be governed by the gravity in selling tar. The lighter in body the less value it has, and 10 per cent. crude oil could not be called tar, and if 10 per cent. crude oil was demanded from me on a tar contract, I would not admit my liability to deliver it; but as to whether I would deliver it, would depend greatly upon the price at the time. If I should tender a tar 8 to 10 per cent. crude oil and 20 gravity up, I would say that the defendants would be right in rejecting it, as it would be unsuitable for making paraffine oil. I would consider it was sold subject to the defendant's inspection, and we have always acknowledged such inspection for coke, water and gravity.

Cross-examined by plaintiffs: If I had been making 10 per cent. tar two years previous to this contract, and the defendants knew it and had been receiving it, I would consider I was right in tendering it to them. They inspect all that is for shipment, and to such inspection as is customary from the defendants we would accept. I have refined petroleum West, and there is no general percentage for tar there, as it is all governed by the demand, but should say the average West is more than 5 per cent. crude oil. (Inspector's certificates shown witness marked Exhibit No. 4.) On these certificates I would say that according to the gravity where it is noted on them that it is all too high to apply on this contract, and we have had residuum rejected for coke and light gravity. Residuum is the heavy refuse left in the stills after the distillation of crude oil. If I had taken from you 200 barrels 10 per cent. tar for experiment and had rejected it. I would say that was sufficient notice to you that I did not wish any more. If I had always been making 10 per cent. tar, and had made contracts, and tar went up instead of down, I would not feel justified in running it down to 4 per cent., but would deliver the same percentage as usual.

Examined by the committee: I would say that if crude oil declines we leave more residuum in the stills so as to save our still bottoms. If residuum is worth eight cents and crude oil five cents we would try to get rid of as much residuum as we could, but if crude oil was higher, as it has been previous to this, we would run it down as low as possible, when it would be more suitable for making paraffine oil. We have contracts with Libby & Clark, the same as plaintiffs have.

Mr. Rockefellars' statement: I am a refiner of petroleum, and our capacity is about 1,500 barrels a day. Petroleum tar is 3 to 5 per cent. crude oil, and 18 to 20 degrees spec'fic gravity. (Contract shown.) On that contract, which is just the same as we have with the defendants, I would not, under these certificates shown, consider it a proper tender, and the defendants would be right in rejecting it, as 10 per cent. crude oil and 25 gravity is not tar, and in fact I never heard of 10 per cent. tar until this case came up. If it was tendered me under those certificates on that contract, and on same certificates the gravity is not noted, which is the most important point, and what I would want to know about, as I would not accept it if over 20 gravity. There is no established rule, although we always inspect for gravity, and above 20 and over 5 per cent. I do not consider to be a merchantable tar.

Cross-examined by the plaintiffs: If we had always made a 10 per cent. tar, and tendered this, and they had received it, I should think they had no right to reject this. We have never made any residuum at our refinery in Cleveland, and have never sold any West, as we use the whole product there ourselves.

By Jas. Donald: (Affidavit put in evidence. Exhibit No. 6. Contracts and certificate shown witness.) I think on that contract they would be justified in rejecting what has been tendered according to the certificates shown. In my own case I would not expect it to exceed 5 per cent crude oil, but if 10 per cent. was demanded, I would tender it if it paid me best.

Cross-examined by plaintiff: If it had been my custom to make 10 per cent tar, and if they had been in the habit of accepting it, I would consider this a fair tender on that contract.

If having made a contract for all my tar, I would not change my system of making it to suit a certain purpose of defendants, if I had been delivering the same kind previously, and they had not objected to it.

By the plaintiff: The defendants ship residuum, and when they made this contract they had all the residuum bought up. Cargoes have always been sold previous without any regard to the gravity, and under these inspectors' certificates we claim it is right as tendered: as they are the recognized inspectors of petroleum and its products, and previous to this it was only inspected for coke and water.

By Atchison Scott: Have been in the business of refining petroleum three years in Brooklyn, and at one time was an inspector for a manufactory, but never inspected for shipping (affidavit put in evidence, exhibit No. 7), 22 to 26 gravity is not tar but oil, and it could not be used for making paraffine oil. (Contracts and certificates shown). On these certificates it is above the ordinary gravity of tar, which should be from 3 to 5 per cent.

crude oil and 16 to 20 gravity. The gravity has never been taken until a short time ago, and I seldom heard it mentioned until this case came up, and as an inspector, judged from my knowledge of it, and inspected it free from coke and water.

The higher gravity the less heavy matter you have, and tar from 22 to 26 gravity I would not accept on that contract.

Cross-examined by plaintiff.—I have heard of residuum being used for making paraffine oil and wax, and a small percentage is used for making gas, and small refiners make heavy oil from it.

By A. T. Smithes.—(Affidavit put in evidence marked exhibit No. 8.) I have refined petroleum, and I define tar to be not above 5 per cent. crude oil, and from 16 to 20 gravity. We never run tar 22 to 26 gravity, and it would not pay to buy 22 to 26 gravity tar to make lubricating oil from. We inspected residuum some times at the factory, but not for gravity, which was some years ago. We sold it to be a merchantable article free from coke and water, and when it was above 20 gravity and parties objected we would make it right with them, and when it is from 24 to 25 gravity would consider it an oil, and would not accept it on contract as it should not be over 20 gravity.

Cross-examined by Plaintiffs—Tar is used for gas purposes, and for that purpose light gravity tar is most valuable, and also for making burning oil. Heavy tar or under 20 gravity is used for making lubricating oil.

At this stage of the proceedings the meeting adjourned until Wednesday, December 17, 1873, at 3 P.M.

W. E. FLETCHER,
Secretary.

NEW YORK, December 17th, 1873.

Adjourned meeting of the Arbitration Committee held this day at 3 P. M., to hear the case of Messrs. Sone & Fleming *vs.* Libby & Clark. All the committee present. Mr. Sherwood in the chair.

The defendants stated that there was no difference between the plaintiffs and themselves as to the gravity of this tar, and would impress upon the minds of the committee that the inspectors named are not experts. I will present as a witness Mr. R. W. Burke.

By Mr. Burke: I am a petroleum refiner, and have been for the past ten or twelve years running three refineries (contracts and certificates shown), under my contract with defendants, which is the same as the plaintiffs. I have always delivered what is customary for me to sell. (Letter put in evidence marked Exhibit No. 9, and signature acknowledged by witness). I have always made the percentage not over three or four per cent, and have delivered that percentage to defendants, and would not feel myself authorized to deliver them a larger percentage. There is no rule in regard to it, but, with one exception, we all refine it down to less than five per cent. Gravity of tar was first brought to my notice about two years ago, occasioned by the examination of some Canada tar, which was about 20 gravity, and eight to ten per cent. Tar of 22 to 26 gravity would not be a fair tender. The term "subject to inspection" means subject to buyer's inspection, if he demands it. If ten per cent tar was demanded of me on a contract I would not deliver it unless I chose to.

Cross-examined by the plaintiff: I have refined petroleum for four years in this State, and I do not know whether 22 to 26 gravity tar is merchantable or not, as I have never made any, and do not know the demand for it, and have never sold any cargoes of tar, neither do I know the value of 22 to 26 gravity tar. I have heard of tar being used for making lubricating oil, but for no other purpose. I have never drawn any ten per cent tar from my stills, and have never delivered tar of 25 gravity to the defendants. I will correct my former statement by saying that I have sold tar for the purpose of manufacturing gas. If I had sold 1,500 barrels of tar. my option, 22 to 26 gravity, I would deliver the heavy. With crude oil at ten cents a gallon, I would not deliver 26 gravity tar at as low a price as I would 22. With crude oil at $6\frac{3}{4}$ cents, I could not say what I would do, as I have made no computation. The value of still-room for crude oil from March to October, 1873, was considerable, but cannot tell its exact valuation. I have never heard the gravity of tar stipulated on cargoes previous to May, 1873, except in the case of the Canada tar mentioned, when the government stopped it from being imported as tar. I run about 8,000 to 9,000 barrels of crude oil a week, making two to three per cent of tar.

By A. H. Mumby: I have been a broker in petroleum and its products for years. Our firm is Mumby & Beckwith. (Contracts shown). I made those contracts, and supposed I was buying the same as I had from other refiners under the same name. I do not know what petroleum tar is, but from hearsay I understand the gravity to be from 16 to 20. The words "subject to inspection" mean a merchantable tar, subject to the buyer's inspection. I did not know how much tar would be delivered. The seller may have said something at the time as to the quantity to be delivered, but I had the impression it would be large. judging by the size of the refinery. I have not heard of any one selling large quantities of petroleum tar for account of plaintiffs previous to this contract.

Cross-examined by plaintiffs: I forget as to whether you said anything to me about the quantity, but my impression was that it would be large. but nothing was said about quality or gravity.

Examined by the committee: Nothing was said at time of purchase that it was intended for any special use, and I never knew of any difference in quality between this refinery and others. Defendants are large buyers, and they sometimes give us orders to buy right along at a certain price, but in this case I do not recollect if they gave us the order before I saw the plaintiffs or not, but I had no particular instructions. As a broker I was trying to make a sale.

W. N. Day's statement. I have been interested in petroleum and its products for some time. (Contracts shown.) I should say that that contract called for ordinary merchantable tar as made here.

Examined by the committee: Merchantable tar I judge should not exceed five per cent. and any greater percentage should not be sold as tar. From 22 to 26 gravity, and eight to ten per cent crude oil, I should call an oil, and it would not apply on that contract. They would be justified in tendering it if the buyer would accept it. As to the buyer being bound to receive it I could not say, as there might be such a thing as eight to ten per cent tar. An ordinary crude oil that yielded ten per cent would be a very large percentage, and I could not call it tar.

By defendants: Petroleum tar has always been inspected for coke, water and gravity. The essential requirements are weight and body. We purchased from plaintiffs about two years ago 200 barrels of what was termed ten per cent tar, and it was represented to us as being more valuable for our business, and we paid two cents a gallon more for it. We took it to our refinery and got from it a light volatile oil that made it totally unfit for our business, and we reported it to plaintiffs, and up to the time of this contract we had received no more. We could not use it at our works without our knowledge, as it would not give the yield of wax or heavy oil. The first one or two deliveries from the plaintiffs were sent to our refinery and put in our tar tank, and from there into our stills. We were always testing the gravity and found the oil lighter than what we had received before, and not suitable for our purpose, and it stood in the tar tank at 24 instead of 18, as we had received for the past five years. I did not know where it came from, and inquired on 'Change, and submitted it to two or three refiners, and they said they were not making light gravity tar. I saw the plaintiffs, and they said it was the best tar and suitable for our use, and that it was all right. After this we had each particular lot inspected for the gravity, and then ascertained it came from them, and informed them of it, and asked them to make it a merchantable tar, and we received one or two other lots under protest. We never received anything similar from plaintiffs or any other refiner, because we could have detected it at once, and we could not have used it. Coke and water are no characteristics of tar. That clause only gives the buyer the privilege of rejecting it if it contains coke and water. Mr. Purdy has inspected our tar for the past five years, and no inspection of ours has been questioned.. We have rejected it for body also for containing benzine and naphtha. We make heavy lubricating oil, and the heavier the tar the heavier the oil, and we rejected this tar for light gravity. Before this contract we did not inspect each lot, but since then we have made a gravity of our own, which we do to show the difference in quality. Previous to this contract shippers were not in the habit of buying by gravity because there was no such tar as plaintiffs on the market, which we say is a new article. We have bought the products of refineries at the request of the refiners, as they said their yield of tar was so small that they could not tell what it would be.

Cross-examined by plaintiff: We sold 500 barrels of tar to Ackermann, and put 500 on board the " Eden," but do not know who inspected it. I think we have shipped tar without a certificate. The running capacity of our works now is from ten to eleven hundred barrels a week, and we are receiving about 1,000 barrels a week from refiners. I have conducted the business of making lubricating oil, but do not say that I can tell the gravity of tar by looking at it; but by puting a stick in it, I could tell in that way if it was lighter. I do not know as I could tell the difference between 22 and 24.

Examined by the committee: The ten per cent tar when sold to us was represented to be an article different from others, and better for our business. A few months after this purchase and rejection we bought four or five lots from them, and it was different in character, and never knew they were running ten per cent tar until we discovered it at our refinery. I cannot say how long it was after the receipt of it before we objected to it. We have tested the gravity of tar occasionally previous to this contract. We discovered this article about a week after the date of this contract, and at the time the contract was made we knew nothing of its gravity.

Re-cross-examined by plaintiff: We notified you that the tar did not suit about January 1st, and told you I thought you had a good joke on us by making us pay two cents a gallon more for it when it was worth less.

The defendants here rested their case with the right to call another witness, and also the right to put in documentary evidence.

The plaintiffs presented as their first witness Mr. Lounesbury who stated as follows in answer to examination by the plaintiff: I have been acquainted with your system of manufacture in a general way, and know that two years ago you commenced to run your stills quick, and know that you have continued to do so.

By Paul Babcock, jr.—We have bought tar for shipment for a number of years, and have never heard the gravity of tar mentioned previous to May, 1873. In buying we consider the lighter the more valuable, and always avoid buying a thick heavy tar. I have an order for 26 gravity tar, as shipper said heavy gravity tar would not do. We bought a cargo of light gravity tar from the plaintiffs on the 25th October at 8 cents for Mr. Simmonds, an English shipper (as per Exhibit No. 11 put in evidence).

I consider light tar a merchantable tar strictly prime. There were but two sellers of tar in the market, and they controlled it all, and about eight or nine months ago we tried to get some but could not.

Cross-examined by defendants.—I understood you had contracts with all refiners up to January 1. We had a large order, and could not get the tar from plaintiffs, and we heard Lombard, Ayres & Co., had the kind of tar we wished, and they said they would deliver to us if you would give your permission, and I offered 9 cents for it. (Contract shown.) I should say that that called for all the tar made at that refinery which is the residuum left in the stills after the distillation. (Certificates shown.) Question asked if they would apply on that contract. Ans.—Purdy's does not specify coke and water which is necessary. Lockwood Bros. & Holly would apply. Edward Harrison's would apply. J. B. Miller's would apply. We believe tar is used largely by gas companies, and also by lubricating oil companies. We should call petroleum tar a heavy residuum left after distillation, which knowledge I obtained from having been in the business of refining oil. The percentage differs according to manufacture; as I made it, it was about 8 per cent. I would consider an article with 50 per cent of burning oil in it as petroleum tar.

I have not sold any tar as mentioned previous to this year. We had not sold a cargo prior to this sale, but have tried to buy cargoes of that character for two years past, we did not buy by gravity, but because we thought it was the lightest in the market, and our buyers wanted a thin liquid tar. I have not heard of tar being rejected on the other side on account of heavy gravity.

Light tar has been rejected on account of heavy naphtha in it.

Examined by the Committee.—I have bought tar in this market for 8 years. The exportation has varied a great deal. The defendants by their operations have decreased it. I have no idea of the proportions of exports to the production of tar.

Re-Cross examination by the defendants.—I think prior to this year there has been an average of two cargoes a year exported from this market. The amount of tar required for shipping creates an open market with a regular demand for it.

Re-direct examination by the plaintiffs.—It was a well known fact for the past two years that your tar was the best in the market.

By Mr. Lawrie.—I have bought tar from the plaintiffs and considered I was buying a strictly merchantable article. (Certificates shown). I should say they called for tar or residuum free from coke and water. I have known of your making light gravity tar for about 2 years, having bought a cargo from you some two years ago. My orders when buying tar were not to have it heavier than 16, and our instruction to inspectors was to be free from coke and water, and no limit as to light gravity.

Cross-examined by the defendants.—I think at the time the cargo was to be of a liquid quality. The plaintiffs told me that the lighter the gravity the more valuable it was.

I stated at a meeting of the petroleum trade that the lighter the tar the better for export. I wanted the gravity from 16 to 20, and not so low as the defendants wanted it.

I do not think now that for certain purposes the light tar is the most valuable. I bought a cargo of you for the Eden, did not know who made it. No gravity was asked; it was only to be a merchantable tar; it was shipped to Liverpool, and we did not hear anything about it until Mr. Kimball, partner of defendants, went there last summer, when we heard complaint. I am satisfied with the inspection of Lockwood Bros. & Holly, and I have purchased cargoes in Philadelphia, and accepted them on such certificates.

I have made a reclamation on such certificates before this committee, and if it had been from 16 to 20 gravity do not know as we would have had the trouble. I purchased a cargo from plaintiffs in September without the gravity being mentioned, but I supposed it to be about 26, and that cargo was accepted on the other side. I received orders afterwards for heavy gravity, and bought it from defendants.

Re-direct examination by the plaintiffs.—I have seen copies of the circular issued by Libby & Clark on the other side, which speak unfavorably of light gravity tar.

By Mr. Bunker.—I have known that the plaintiffs have been making a light gravity tar for about two years, and they have claimed it to be superior for that reason.

I have purchased this tar from them and think it to be a superior and merchantable article, and have not had any complaint. (Contracts and certificates shown.) I should consider the certificates correct. I would have given as good a price for plaintiffs' tar as for any other until recently, since the orders came for gravity.

Cross-examined by the defendants.—I shipped tar in 1872, but did not know the gravity. I do not think I ever said a merchantable tar should be from 16 to 21 gravity.

Examined by the Committee.—I have bought tar for lubricating and gas purposes, and always to ship, but never examined it.

By Mr. G. F. Gregory.—I have refined oil about 5 years. I have never bought any tar, and never sold any except my own, and the highest I have ever delivered was 3½ per cent.

I have never heard gravity mentioned previous to May, 1873, and I do not think the gravity of tar could be told by looking at it.

The certificates shown are all valid in authority; 26 gravity tar is more valuable than 18.

Cross-examined by defendants.—Our tar runs from 3 to 4 per cent. I do not know as I would be right in asking you to take 10 per cent crude oil, and do not think I could be compelled to deliver 10 per cent. We say light tar is more valuable because it has more burning oil in it. I do not know how valuable it is for lubricating oil, as I never made any.

Examined by the Committee.—I could not say what tar is meant on that contract, and it would depend upon the orders in the market at the time. If it contained 10 per cent crude oil it would be tar, but we never make it more than 4 per cent. The tar would depend upon the kind the plaintiffs were making previous to the contract, and if they had been making it for a year previous it would apply. Early in the fall I was told by the plaintiffs' foreman that they were making light gravity tar.

By Mr. McGoey.—Examined by Plaintiff.—I have heard that you were running from stills quick, and believe it is so, but have not seen it done.

Cross-examined.—I have been a refiner 15 to 18 years, and there is a great difference in the running of the stills; we run them close. I would refuse to deliver 10 per cent., and could not be compelled to.

On motion, the meeting adjourned to December 19, 1873, at 3 P. M.

WM. E. FLETCHER,
Secretary.

———

NEW YORK, December 19, 1873.

Adjourned meeting of the Arbitration Committee held this day at 3 P. M., to hear the case of Sone & Fleming *vs.* Libby & Clark. All the committee present. Mr. Sherwood in the chair.

The plaintiffs called as a witness Mr. Abm. McCreery, their superintendent.

By Mr. Abm. McCreery: I am superintendent for Messrs. Sone & Fleming, and have been ever since they commenced refining. I have refined petroleum about twelve years, and consider myself an expert in running and building a refinery. We have been running quick for over a year prior to March 8, 1873, three or three and a-half times in the seven days. I have kept a record in a book of the time the stills start, and the time they are off. We have produced about ten per cent of tar during the past twenty months, and our tar yields about eighty per cent of oil. There has been no change in our running for the year previous to March, 1873, and I would have known of any change had it occurred. The lighter the gravity the more valuable for general purposes. It is used principally for making illuminating oil, also lubricating oil, paraffine wax, gas and aniline colors. I first heard of the gravity of tar last April. We have not been idle during the past year at our refinery.

Cross-examined by defendants. Petroleum tar yields about eighty per cent of burning oil, of which the best is run into burning oil, and a portion of it would make a heavy lubricating oil, about twenty per cent of it.

Although I have not been in the lubricating oil business, I know the principle of it. I have tested the gravity at different times, but never found it above 25 nor below 23.

Examined by the committee: Tar has never been of uniform gravity. We used to make it heavier prior to two years before this contract. I did not try the gravity, but judged it would be heavier by the quantity. The gravities made in different refineries vary, as some have more tar in their stills than others, and the refineries that I have worked in have averaged about five per cent.

Re-cross-examined: About twenty-two months ago we commenced to run ten per cent tar; previous to this it was about seven to eight per cent, and 22 to 23 gravity. The first twelve months that we made twelve per cent tar, we sold some cargoes prior to this contract, I think, but do not know its gravity. We did not endeavor to keep the making of it a secret, as we had no object in doing so. In 1872 we sold some to domestic buyers. It had some coke and water in it, for which allowance was made. It was sold to the defendants. It came from the bottom of the stills, and was dipped out by the still cleaners, and it contained about thirty-five to forty per cent coke. About ten per cent coke is left after we run off ten per cent tar. The coke that was in the tar that was scooped out was soluble coke. I do not know as the tar would be any heavier. The coke might make it show heavier, but I never took the gravity of it. The variation in the gravity is on account of our stills not being covered, and when we have a rain storm it would keep them back, and there would be a few more barrels of tar in the stills, and the more tar left, the lighter the body, and the gravity would be according to the quantity, about 25 to 26. The percentage of heavy oil in ten per cent tar is about thirty per cent.

J. Lombard, Jr.'s, statement: I am an oil refiner, and we made seven to eight per cent tar when we run our huge still, which was all our refinery except our tar stills. I never thought of the gravity, but considered it a merchantable tar. We have delivered some of that tar to defendants at following times: December 5, 1871, 79 barrels, and during the year 1872, 1.700 barrels of the same tar, and there was never any complaint made direct to us by defendants prior to complaint on this contract. We made other sales, 400 barrels to R W Burke, 300 to Stuben & Manger, and no complaint was made of either. The latter party I believe used it for making illuminating oil. The gross amount sold in 1871 was 1,480 barrels, and in 1872 over 2,000 barrels. The plaintiffs told about running their stills quick about two years ago.

Cross-examined by defendants: The crude oil used when we made seven per cent tar was not heavier than usual at that time, but it is heavier than what is used now, and eight per cent of it would be heavier then than it is now. I never tried the gravity. A delivery of 400 barrels made to you was put on a schooner for Boston, which was part of a lot of 700 barrels sold you. We sold you 500 more delivered in four lots, and 500 more delivered to you in four lots. They were from crude oil of a gravity of 45. (Letter put in evidence with certificate attached, Ex. No. 12, dated August 12, 1873, signed Lombard, Ayres & Co.; also one dated November 29, 1873, Ex. No. 13). All crude run heavier when we were making those deliveries than now. The tar would be heavier, but how much I could not say. It would not be of as light gravity as ten per cent of the crude oil is now. (Contract shown and witness examined by the committee.) We should deliver on that contract the same tar as we were making at that time. (Certificates shown.) I should consider the tar that we had been making a proper tender on that contract, if they had been receiving the same kind previous. I never knew of any rate of gravity at the time. We run our stills about the same as the plaintiffs, and have no regular percentage in the stills. We run down our large stills to a low figure, but not so as to burn our still bottoms, and when it did not pay to sell tar we put it in small stills and made refined oil out of it. We would call it tar when we put it in tar stills, which were only used for running tar, and we considered it a merchantable article when we put it in our tar still, it being seven to eight per cent of the large still, which we could not run down any closer on account of its shape. Previous to this year we paid no attention to gravity, and, when we did so, it was on account of hearing that the defendants would not take any tar from us over 22 gravity; but we did not do it, because we thought the tar delivered previously was not merchantable. When we put tar in the tar stills we run it down to coke, which is all that remains.

Re-direct examined : We made seven or eight per cent tar; you made ten per cent, and all others made about five per cent. There was no rule as regards the percentage; it was all a matter of profit.

Re-cross-examined by the defendant: I would not consider I had a right to tender fifteen per cent on a tar contract, but it should be governed by the trade previously in New York, which custom was to deliver five per cent, with exception of the plaintiffs and ourselves.

The plaintiffs here presented and read a number of affidavits of employees marked Exhibits Nos. 14, 15, 16, 17, 18, 19, 20 and 21. Letter from Phœnix Petroleum Works, signed Malcolm Lloyd. Ex. No. 22. Letter from Imperial Refining Co., Oil City, signed by John Gracie. Ex. No. 23. From T. A. Allen, Corry, Pa. Ex. No. 24. Contracts with Robert Norman. Ex. Nos. 25 and 26. Letter from Sinclaire & Marvin. Ex. No. 27. Still book put in evidence showing when they are put on and when they are taken off. Ex. No. 28.

By the defendant. The letter presented by the plaintiff from the Downer Kerosene Oil Co. is not from the manager but from the bookkeeper, as I will prove by a letter from Mr. F. Habershaw, the New York manager (put in evidence), also a letter from Samuel Downer himself marked Ex. No. 29. (Affidavit of F. Habershaw put in evidence. Ex. No. 30. Letter from Hoft's & Co. to Robert Norman put in evidence. Ex. No. 31.)

L. V. Louis' statement.—Our capacity varies from 8 to 10 thousand bbls a week, sometimes it goes over 10 thousand. It was the same at the time of making contract as it is now with the exception of 8 new stills that started the day before yesterday. I stated that we ran one-third of the crude oil at one time that was run in this market, and at the time of the making of this contract we did, but there are two refineries now nearly as large as ours is. We began to run 22 per cent tar about 22 months ago, previous to that time our tar was like most other refiners as we had small stills. (Statement was here presented by defendants showing their receipts from plaintiff since they were in business). We did not take the gravity of tar previous to this contract. In 1872 our tar was light.

I could not tell you the difference in lots of tar made as I did not take the gravity. I think the lighter the tar the more valuable for making lubricating oil under a proper system.

The heavier the tar the better for making heavy lubricating oil, but there would be more less in coke, and I believe light tar is the best for your purpose. I consider that the 60 per cent of oil which you obtain from distilling our tar to be worth almost as much as crude oil, and I paid defendants nearly the same price for all that was offered us, giving them $3\frac{1}{2}$ cents a gallon for it, when crude oil was worth 5 cents.

We are willing to pay that price even if other refiners refuse it, and can prove by our books that it is worth it to us. We always take heavy oil and heavy naphtha together at crude oil price. (Defendants here presented a sketch representing a section of a still. Exhibit No. 32.)

We transfer our 10 per cent tar direct to the tank, and pump it from there; it is a Western process, and we never endeavored to keep it a secret and the lots spoken of as scooped out were taken out of the still below the tar pipe. In 1872 we utilized our tar. We sold 2,000 bbls from the time we commenced to make 10 per cent tar up to the contract which includes what was delivered to defendants, about 550 bbls.

We sold to Sinclaire & Marvin 600 bbls, and delivered it on February 7 and 10, 1872; 1,000 gallons to R. W. Burke; 10,000 gallons to defendants on February 12; 750 bbls to Sinclaire & Marvin on March 14. We delivered to Mr. Baily one of the four lots that were taken by defendants. When it is to our interest we make illuminating oil from our tar, and the reason that Mr. Babcock, the former witness, did not get any tar was because he would not give us money enough for it. We could not leave 15 per cent in our stills if we wanted to, as we would then have refined oil in. From

October, 1872, to October, 1873, we made about 40,000 bbls of this tar from which we made burning oil. The difference in the gravity of the tar is on account of the oil running freer at one time than at another, and the stills will run closer, and we will have storms that will keep the stills back, and there will be more tar left in the stills, and the average percentage will vary about 2 per cent. We have drawn off tar of about 14 to 15 per cent crude oil. but we would not be justified in sending it on a contract, and would not. About four-fifths of refiners outside of New York make 6 to 10 per cent tar.

We do not repudiate Mr. Purdy's inspection, and do not know as we ever had any 22 gravity tar, but if we did it might be 6 or 7 per cent as well as 10. We might have 100 to 250 bbls 22 gravity come from a still. In 1872 we might have made 50 to 70 bbls of 22 gravity ; it might be more, as it is impossible to get a uniform gravity of tar. I would prefer to sell the heavy gravity tar, as it would not make illuminating oil. I do not think the deliveries in 1872 could all have been like the lot of 22 gravity delivered on May 19. The tar is no heavier, only the coke gets mixed with it and it is slushy.

I do not know the gravity of the 150 bbls spoken of as delivered in 1872.

I have not tried to influence any refiners to make this tar, and have made no statements to any one that defendants were violating their contracts with other refiners, and have not heard of it before. I have not tried to influence a single broker by promising to give them business.

I have offered you heavy oils on one or two occasions, and might have been in your office, but do not remember of hearing you say that the 10 per cent tar was unsuitable for your purpose.

Examined by the plaintiff.—I could have sold this tar for shipment at the time we made contract with defendants, but we did not for the reason that we might avoid delay in procuring vessels, and also in selling in product as in case of a fire our deliveries would cease and we could have obtained one cent a gallon more for cargoes.

Examined by the committee.—The broker came to me and asked me if I would sell tar, but nothing was said about gravity or quality, nor any reference made to the 10 per cent tar sold previously, and he did not ask what percentage we were making.

He gave the name of the buyer, but did not say what it was to be used for. The only talk we had was about the return of the empty barrels, but nothing else was said different from the contract.

Cross-examined by the defendant.—We could not run our stills down close, and deliver you a smaller quantity. If we transferred our 10 per cent tar to our small stills and distill off 50 per cent, it would not be the production but would be pitch.

Examined by the committee.—If I distill 10 per cent tar and take oil

50 per cent. 7 to 8 per cent would be gis, and the oil would be nearly worth the price of crude oil or the same as defendants paid us 3½ cents a gallon for. We sold them the 200 bbls of what they call 10 per cent tar, but do not know as I told them it was our product at that time but the trade all knew it. When we sold them the 200 bbls we were building our tar stills and could not use it, and felt as if we were sacrificing it in selling it, but do not know as we told defendants this, but we mentioned it to several buyers. We claimed a specialty for it, and it was known to the trade previous to this contract.

Cross-examined by defendants.—I do not at the time I sold you the 200 bbls tar recollect that I represented that it contained more oil and was better for your business. The 2 cents a gallon extra price was charged on account of its light gravity.

Re-direct by plaintiff.—A proposition was made to me during the existence of this contract to settle it for $1,000, which offer was in June, a.d I refused it, as we had sustained more damage. I was requested to call on defendants a number of times, and I finally said that if they would pay us $6,000, which was less than our damage, and receive our tar to September 1st, I would cancel balance of contract.

By defendants.—We admit having one or two interviews in regard to a settlement of this matter. There was no settlement made, and we did not make any such offer as plaintiffs say. They made us an offer which we did not accept, and everything that passed at that meeting was to be kept secret, and not to be mentioned in this case.

At this stage of the proceedings the case was adjourned to Monday, December 22, 1873, at 3 P. M.

WM. E. FLETCHER,
Secretary.

NEW YORK, December 22, 1873.

Adjourned meeting of the Arbitration Committee held this day at 3 P. M. Case of Sone & Fleming vs. Libby & Clark. All the Committee present.

By the defendant: I will now offer in evidence a letter received from. Mr. Lombard since the last meeting in regard to the question I asked him, if it would not be possible that the tar delivered us in 1871-2 from 7 per cent crude oil might not have been 22 gravity. (Letter read dated 22d December, Ex. 33. Letter from Lombard, Ayres & Co., dated December 19, put in evidence, Ex. No. 34.) I will also present an affidavit (Ex. No. 35) from Wm. O. Allison, Esq., editor of the *Oil, Paint and Drug Reporter*, with schedules marked A and B, showing the export of residuum, and in explanation I will say that in 1871 there were 4,447 barrels exported, and in 1872 6,980 barrels. In 1873, up to December 1,

14,900 barrels from this port. Our contract was made with plaintiffs on 28th March, and by noting the figures contained in this statement it will be seen that we could not have possibly bought tar for export instead of for use. I will put in evidence letter from Samuel Downer, of Boston, dated December 2, 1873, Ex. No. 36. Letter from Mr. Samuel Downer to Mr. Habershaw, December 21, Ex. No. 37. I would say that Mr. Downer is the pioneer of the petroleum trade in this country. I will offer in evidence our correspondence, which will show that we protested against this tar as soon as we discovered what it was, and that we were willing at any time to receive a merchantable article. (Letter from Libby & Clark to Sone & Fleming, April 1, put in evidence. Ex. No. 38.) The protests previous to this were verbal. (Letter from Libby & Clark to Sone & Fleming, May 19. Ex. No. 39. Reply from Sone & Fleming, May 19, Ex. No. 40, which is their first reply. Reply by Libby & Clark, May 20, Ex. No. 41, showing that they, Libby & Clark, offered then to submit the case to arbitration on 500 barrels delivered. Letter from Sone & Fleming, May 24, tendering 300 barrels. Ex. No. 42. Letter from Libby & Clark, May 24, refusing tender. Ex. No. 43. From Sone & Fleming to Libby & Clark, May 27. Ex. No. 44. From Libby & Clark to Sone & Fleming, May 28. Ex. No. 45. No answer was received to this. Copy of an order to Mr. Purdy to inspect 1,000 barrels; also letter indorsed from Libby & Clark to Sone & Fleming, upon which we have our reduction. Ex. No. 46.) We received delivery orders every day for the same lots of goods as we had previously rejected. (Order June 23 for 1,000 barrels and reply by Libby & Clark. Ex. No. 47. Order for 1,000 barrels June 30, and reply attached from Libby & Clark. Ex. No. 48. Invoice from Sone & Fleming, with reply from Libby & Clark attached of July 7. Ex. No. 49. First notice received of any petroleum tar that was sold for our account dated July 28. Ex. No. 50.) The notice of sale gives no names, and we think it was to try and show that tar was not worth more up to July 28 than the contract price, and we will show that it was worth more. Order received for 1,400 barrels about this time. to which we replied by letter dated 29th July. Ex. No. 51. Order on September 3 for 1,500 barrels, with our reply dated September 6. Ex. No. 52. We sent an inspector to their yard upon receipt of every notice, and will present the inspector's affidavit very soon. Order received September 13 for 1,000 barrels, and reply sent to Sone & Fleming. Ex. No. 53. We did not care to send an inspector to examine the same goods over and over again, so we sent our young man and asked if it was the same lots, and they said they did not know, but we could send and find out, as they had no time to write letters. We received another tender on September 17 for 1,200 barrels, and sent reply attached marked Ex. No. 54. Notice of second sale received on September 17 for 1,800 to 2,500 barrels. Ex. No. 55. Order for delivery of 2,000 barrels, October 11, and our reply attached. Ex. No.

56. Notice of sale of 550 barrels, October 11. Ex. No. 57. Notice of sale of 500 barrels, October 16. Ex. No. 58. We received frequent notices to go up and inspect tar. On October 28 we received notice and replied October 29. Ex. No. 59. From Sone & Fleming to Libby & Clark, November 15. Ex. 60. Order for 2,000 barrels November 15. Ex. No. 61. Statement of a sale on October 25. Ex. No. 62. Statement presented showing all the sales made for our account. Ex. No. 63. Letter from Sone & Fleming, with paper attached November 15. Ex. No. 64. Mr. Purdy, when he found any additional tar, he put it on certificate. (Affidavit of Robert Stewart, with memorandum attached, dated December 11, Ex. No. 65.) The first tender of tar was made before May 17. We will put in evidence statement from Custom House at Buffalo, dated June 26, Ex. No. 66, to show that all above 20 gravity is rated as oil and all below 20 as tar. There is a specific duty of 20 cents a gallon on oil and an ad valorem duty of 20 per cent on tar. The business in illuminating oil was large at that time, and it was to make the duty on it prohibitory, but it does not indicate the value. We introduce this to show the Government standard, and it has reference to Canada tar. (Affidavit of T. T. Parsons put in evidence. Ex. No. 67. Statement read. Ex. No. 68, showing that defendants have kept their contracts in good faith, signed by all the refiners in New York. When we found out the gravity of the tar we were receiving from plaintiffs it was pumped from the tar tank and sent out of store. (Affidavit of John Plunkitt read, marked Ex. No. 69. Letter from James Goldsmith read. Ex. No. 70. Affidavit of George Sommers. Ex. No. 71. Affidavit of Wm. A. Byers. Ex. No. 72. Affidavit of R. T. Burt. Ex. No. 73. Letter from C. Huron & Co. Ex. No. 74. Affidavit of S. Jenny, Jr. Ex. No. 75. Affidavit of E. G. Kelly.) These affidavits and letters, with the testimony of the witnesses we have presented, is the expression of every producer of tar in New York and vicinity.

Mr. H. H. Rogers, of the firm of Chas. Pratt, was here recalled by the plaintiff, and stated as follows:—If defendants had been receiving this tar previously in quantities they would be bound to receive it on this contract.

Cross-examined by defendants.—If you had been receiving this 10 per cent tar previous to this contract without protest you would be bound to accept it on this contract. I do not think they had a right to tender 10 per cent tar on this contract, but if you had been receiving it and protesting I do not think the plaintiffs would have a right to continue to tender it. If 10 per cent tar was put in small stills, and 50 per cent distilled off that remaining would be tar of about 15 gravity. If plaintiffs had been delivering 10 per cent tar all along, I do not think it would be right for them to run it down if the market went against them. I do not known of any difference between 1st and 2d distillation of tar.

Mr. Clark, one of the defendants, here read his own affidavit marked Exhibit 76, and stated that the plaintiffs had put in evidence letters from the parties west, which left the impression that nine-tenths of the refiners in the United States made 10 per cent tar. I will put in evidence letters from Standard Oil Company (Exhibit No. 77), who run nearly one-third of all the crude oil in the United States.

By plaintiff.---I included the Standard when I spoke of the amount run per day.

By the defendants.—We will put in evidence affidavits from paraffine oil companies in Cleveland (exhibit No. 78), letter from J. B. Merrian (exhibit No. 79), letter from E. S. Thayer & Co., Boston (exhibit No. 80), affidavit Mr. Lincoln, Boston (exhibit No. 81). We are the only parties in New York who make paraffine oil and wax, with one small exception, and we have made in our two factories this year 60,000 bbls. Last year we only had one factory running, and we made from 25,000 to 30,000 bbls.

By the plaintiffs.—We submit our contract which we stand on, and they must say why they have not fulfilled it. I will read letters from Mr. Donners, manager (exhibit No. 25). This man buys the tar and sells the oil and draws the checks, and the defendant has presented a letter which states he was only a bookkeeper. I have telegraphed him, and he replied as follows (exhibit No. 82). I have also received two telegrams stating he is Mr. Donners' manager. (Exhibit No. 73.) I will also read exhibits Nos. 2 and 3 again, also the letter from Mr. Donners' manager (exhibit No. 25). Letters from Mr. Andrews, at Cleveland, put in evidence (exhibit No. 85), and explanation given in regard to the signature.

By Mr. Affley.---I have been in the employ of the plaintiffs about two months, but do not know anything about tar. I had never met any of the gentlemen before who gave me the letters which have been produced here. They all answered my questions, and gave me the letters voluntarily. I saw them all written but Mr. Andrews. Mr. Gracir put the postscript to his letter when I told him he had omitted the percentage. I saw Mr. Allen, of Donner & Co., and he seemed to want to impress upon my mind that defendants and themselves were on most friendly terms as far as their business relations were concerned. Mr. Allen showed me the books where they had made restrictions on tar, where it was below the gravity of 20 deg. I did not see Mr. Andrews, and wrote to him, and asked him to answer my question as concisely as possible, as I could not get to Cleveland until late Saturday night, and left early Sunday morning.

Cross-examined by defendant.---I do not know for what purpose the tar is used by the Donner Oil Co., or how they use it or where they use it.

By defendant.—We withdraw whatever we have stated in regard to Mr. Allen, as Mr. Habershaw must have been wrong.

By plaintiffs.—The defendants in the correspondence did not read the first letter before the rejection (exhibit No. 85). They have never taken away from our yard one-half of their production, and always wanted us to let it remain, and we had to cooper it, and that accounts for the cooperage bill (exhibit No. 86).

By defendants.—We admit that we were tardy in taking it away, and we are under obligations to plaintiffs for the favors granted us by them. I will put in evidence a condensed statement showing all the oil taken in contract (exhibit No. 87).

The plaintiff read a letter dated April 7 (exhibit No. 85). Letter of April 7 from defendant (exhibit No. 88).

By the chairman.—By the statement of defendants it appears that 2,555 bbls. were taken after April 1st, the time when plaintiffs received protest.

By defendant.—I think that 4,355 bbls were received after the contract before positive refusal; but it was all received under protest. The letter of April 1st is the first written protest.

By plaintiff.—About a week before this written protest defendants asked us to make the tar heavier, and we said it was impossible, and that we could not change our sytem of manufacture. After having taken 500 barrels and written they would take no more, they took 2,555 barrels between April 1 and the final rejection. In our statement of loss we propose to allow defendants crude oil price for the heavy oil which is about 80 per cent. We bought 550 barrels from defendants, and we said we would take all the rest they had; we received 800 barrels of it, and gave them 3½c a gallon for it, and it showed by that accumulation that the tar they were running was making some light oil.

By defendant: This tar had been accumulating for some time in our different warehouses, and we did not know how much we had of it until we turned it out. Some of the heavy tar produces some light oil, some about 25 per cent, yours produced 60 per cent.

By plaintiff: The amount of our claim is $32,000 to $33,000 for damages. It is made up by crediting the defendants with tar they have taken and paid for, also with tar sold for their account, also tar run for their account not being able to sell it, allowing them 80 per cent, and giving them crude oil price for it. We debited defendants with total production, and the cost and money expended, also lighterage, cooperage and charges in making the sales. (English publication put in evidence showing the price of our tar. Exhibit No. 89.) We were the largest refiners in New York up to a short time ago, with the exception of Pratt and Rockafeller. A year ago we ran nearly one-third of the crude oil that was run in New York. Our system is different from other refiners here, and we think there is more money in it. We cannot make tar refiners of ourselves, and cannot run our works to suit our waste. (Illustration given by plaintiff to show why they could not comply with defendants request and make tar heavier.) We claim our stills were $1.60 a barrel on all the crude oil we ran, and we tried to run all the crude oil we could. 500 barrels run off a still in 2 days which gives us 250 barrels a day. The 50 barrels of tar left in the still is worth $135, and it would pay us better to let that run in the creek sooner than lose the use of our stills a day. They all know how we were running our stills, and we will prove the value of them at the next meeting.

The committee then adjourned to December 27, 1873, at 3 P.M.

WM. H. FLETCHER, Secretary.

<hr>

NEW YORK, December 27, 1873.

Adjourned meeting of the Arbitration Committee held this day at 3 P. M., to hear further testimony in the case of Sone & Fleming *vs.* Libby & Clark.

The plaintiffs presented additional certificates of inspection of tar covering lots defendants inspected, Ex. No. 90. Letters from shippers stating price paid defendants for tar bought from them, which was accepted by defendants on this contract. One, December 24, from Messrs. Ackerman & Co., Ex. No. 91, and one from Sinclaire & Marvin, December 26, Ex. 92. Affidavit of Mr. C. Page presented, Ex. No. 93. The object of presenting these is to show that they purchased more tar than they could use, and bought ours with the intention of cornering the market, and to also prove that they were shipping tar.

The broker, Mr. A. H. Mumby, was here recalled, and stated as follows:

My partner and I made the contract together, and with the plaintiff, I think. We have to find out who has tar for sale, and go to them and try to buy it. No one asked us to go to plaintiffs. I knew they had it for sale but did not know anything about the quality. I supposed the quantity would be large. I sold it to defendants, and think my partner saw them. I saw the plaintiffs, and he saw defendants. I have been buying tar for years; sometimes we have an order and sometimes we solicit. In this case we found the tar and offered to defendants, and all I recollect in regard to the sale is only what is expressed in the contract. We have bought tar for defendants from the plaintiffs before, and nothing was said about gravity or density, nor percentage of the still, nor reference to any part of the transaction. I knew of no difference between this tar of plaintiffs and other refiners, and expected it to be the same, except a larger quantity, on account of the size of their factory.

Examined by the defendant: I expressed surprise at the time the first tenders were made as to the quantity and the gravity in conversation with Mr. Purdy. At the time of buying this tar I had no conversation as to the gravity, and had no idea I was buying a different article than from other refiners.

Examined by the committee : The gravity of this tar was higher than I expected tar was from what I had heard, but did not think that it did not comply with the contract. I had heard inspectors say that some refiners' tar was as low as 15 and others was as high as 19 to 20. It was new to me at the time, and had been brought to my notice by a lot of tar having been sold here and rejected on account of the gravity. Prior to that time, which was before this contract, about two years ago, I knew nothing of the gravity of tar, and the gravity of the tar then spoken of was Canada tar. The knowledge of this gravity did not effect the trade, as they went on buying as usual without regard to the gravity. There has been preference expressed by customers for different kinds of tar; some liked the Oleophine best, and others expressed their preference for other refiners, as some had more or less coke in it, yet it was all bought as merchantable tar. I have heard from the inspectors that all the refiners ran it from 16 to 20 gravity. They commenced to inspect for gravity at the time this case came up, and have continued to do so ever since.

Examined by defendant : My firm made that contract, which was for merchantable tar, and always when sold to domestic buyers it is specified "subject to inspection."

Examined by the committee : The tar, according to contract, was to be a merchantable tar, free from coke and water ; the percentage and gravity was a new question. As to what constitutes a merchantable tar, I think a refiner's opinion would be better than my own, as I never saw tar run from a still but once, and do not know anything about it. It is always subject to inspection, and sold in that way. I do not know what is merchantable tar now and did not then. I expected plaintiffs to deliver the same as they had before without any reference to any particulars; that has been my habit in making contracts. The buyers, in conversation, have only required clean and nice tar, free from coke and water.

Examined by defendant : I supposed for the past few years that you required a heavy gravity tar.

Examined by committee : They never to me insisted upon having a heavy gravity tar. I knew they had a preference for some grades of tar, for which they would pay a full price.

Examined by defendant : Previous to this contract I never heard of petroleum tar being over 20 gravity.

Examined by plaintiff : Previous to this contract I never heard of the gravity of tar at all, with the exception of the Canada tar, as previously stated. I have never specified gravity at all in any contract either before or since this contract. I do not recollect if Messrs. Lombard, Ayres &

Co.'s tar was light or heavy, as it now is a long time since I have sold any.

By plaintiff: I will offer in evidence contracts for sale of tar for account of defendants to shipping merchants. Ex. No. 92. Also offer as evidence Rule of the Petroleum Trade. Ex. No. 93.

By defendant: I object to the rule as evidence, as it was formed after this contract was signed.

By plaintiff: It was simply a continuation of the old rule, and defendants opposed it; but it was adopted by the trade by a vote of ten to one. (Affidavit of L. V. Sone put in evidence. Ex. No. 94.)

By Mr. Sone: We admit the defendants saw us and asked us to make heavier tar, and we said we could not; that it was impossible under our system of manufacture, and their saying it would not suit for making paraffine oil was not, in our opinion, a protest. (Affidavit of F. E. Fleming presented. Ex. No. 95.) In the letter of defendants of May 19, I would call your attention to " We enclose you inspector's returns for 500 barrels. We accept 150 and reject the balance." They did not take the 150 barrels, and have never sent for it, and further on in the letter say, " We have tried to dispose of what has been tendered us since that date, not a barrel being suitable for our purpose. (Letter of May 19 put in evidence. Ex. No. 39. Inspector's certificates, order bill and guager's return, dated May 19. Ex. No. 96.) We offer also in evidence the minutes of the Complaint Committee and of the Board of Managers. Exs. Nos. 97 and 98. Empty barrels account showing price of empty barrels. Ex. No. 99.) We offer in evidence all of our books for examination by the Committee. Ex. No. 100. (Letter from defendants read dated May 20. Ex. No. 41.) We refused to bring the case of 500 barrels before the Committee to be arbitrated upon, as if they gained that they would gain the whole contract; but if they lost it they would go to law about the balance, and we wanted the whole contract brought before the Committee. When they first rejected it we tried to have it arbitrated, and Mr. Clark agreed to come here and sign the book the following Monday, as his partner was out of town. I signed it, and the following Monday was the first time I had knowledge that they were not going to arbitrate. I put in a complaint to the Complaint Committee, and 23 hours after we received a summons of a suit dated June 12. The case came in August, and we answered it and put in a counter claim, and when the second contract came up we brought a suit against them in Brooklyn, and within a month's time we were notified here by Mr. Fletcher, the Secretary, that they had finally agreed to arbitrate; and you can see that we have been trying all the time to have them arbitrate, and they told the Complaint Committee when they were before them that they would not arbitrate, as there were legal and technical points in it, and their lawyers had advised them to bring it before the courts, and wanted to read a letter from their lawyers, but the Committee would not hear it, and they referred the case to the Board of Managers, where they were again asked to arbitrate, which they refused to do for the same reasons, and were thereupon suspended from membership of this Exchange. When they put it in their lawyers' hands, of course, we had nothing to do but to see our counsel, and we think that their lawyers told them that in the Court of Appeals on one decision there had been ten references. We have been trying to have this matter settled in a business way like business men. I will now offer in evidence three letters from defendants to us, May 19, 20, 24. Exs. Nos. 39, 41, 43. Also balance of stock. Ex. No. 101. We sold this tar with the market at 8⅜ cents for crude oil. After we had been delivering tar the market went up as high as 11 cents, and we were then urging them to take the tar, and they did not take over one-half of it. Exhibit No. 102 will show the price of crude oil, and at the rise of the market we did not cut down our percentage of tar. (Copies of letters presented from plaintiffs to defendants showing how they urged them to take the tar.)

By defendants: Chemical analysis put in evidence marked Exhibit No. 104. Affidavit of R. B. Stewart, Exhibit No. 105. Affidavit of H. Patterson, Exhibit No. 106. We will now put in evidence a bill of sale, Exhibit No. 107, rendered us by Mr. T. C. Baily at the time we purchased the oil works from him on June 25, 1872, two months previous to making this contract, in which bill of sale of 200 barrels of plaintiffs' tar is included, which we will prove by following affidavits was tar from 19 to 22 deg. gravity. Affidavit of T. C. Baily, Exhibit No. 108. Affidavit of John Plunkitt, Exhibit No. 109. We will present affidavits of brokers who have sold domestic tar. Affidavit of W. A. Townsend, Exhibit No. 110. Affidavit of H. N. Curtis and Jeremiah Crowell, Exhibit No. 111. We have testimony from shippers to show that at the time contract was made gravity was not specified, and they said then that light tar was best, but they have changed their minds since. Affidavit of Peter Jones, Exhibit No. 112. Letter from Whitman Bros., Exhibit No. 113. Gravity scale presented showing the difference between this (Am.) scale, and the English scale. Exhibit No. 114. Copy of letter from Henderson, Cooper & Co., Glasgow. Exhibit No. 115. Also a contract made by us with Messrs. Sinclaire & Marvin on the 23d of October, 1873 (exhibit No. 116), when they said light tar was the best, but since then they have changed their minds, and have paid 30 per cent more for a heavy tar, even with the decline in crude oil. I should not say that a chemical analysis was conclusive if we had to depend upon that entirely.

By Mr. Libby: We have now presented all our evidence, and if the committee so desire it I would be pleased to go on the witness stand, and answer all questions they feel disposed to ask.

By plaintiffs: We will present in evidence our claim for damage on the second contract marked Exhibit No. 117, from October 1, 1873, for nine thousand one hundred and fifty-three dollars and fifty-nine cents ($9,153 59); also, balance of our damages to date, December 16 marked Exhibit No. 118, for seven hundred and thirty-one dollars and twenty-five cents ($731 25). We would call the particular attention of the committee to the testimony of ours contained in letters and affidavits from all the leading brokers in this city, and of nine-tenths of the refiners throughout the country. Additional evidence presented marked Exhibit No. 119. I will also read an extract from a letter from F. W. Simonds dated Liverpool, October 14, 1873, Exhibit No. 120. I will read the business card of defendants, on which it says they have cargo lots of residuum for shipment, showing that it was not all bought for manufacturing.

By defendant: That card was made at the time Mr. Kimball went to Europe, which was after we had refused to receive the residuum.

The evidence all being in, the parties stated that they would be ready to sum up on Monday, and on motion of the committee adjourned to meet on Monday, December 29, 1873, at 3 P.M.

WM. E. FLETCHER,
Secretary.

NEW YORK, December 29, 1873.

Adjourned meeting of the Arbitration Committee held this day at 3 P.M. All the committee present. Mr. Sherwood in the chair. The Chairman stated that the meeting was held for the purposes of hearing the summing up of the case of Sone & Fleming vs. Libby & Clark.

Mr. Sone for the plaintiffs summed up his case by reading from manuscript a complete review of it from beginning to end, and concluded by apologising to the committee for the trouble that they had occasioned them and thanked them for their kind attention.

Mr. Libby for the defendants replied to Mr. Sone in a long and exhaustive review, explaining the points of his case in detail and thanked the committee for the attention given in this long and tedious case.

The parties then retired, and the committee after discussing the case agreed to adjourn until Friday, January 2, 1874, at 3 P.M., when they would make their final award.

W. FLETCHER,
Secretary.

The committee, after having discussed the case, made the following unanimous award:

NEW YORK, January 2, 1874.

Whereas, a controversy between Messrs. Sone & Fleming and Messrs. Libby & Clark, members of the New York Produce Exchange, having been voluntarily submitted to the Arbitration Committee of the New Produce Exchange for their decision by an instrument in writing bearing date the seventh day of November, 1873, duly signed and attested, and whereas the proofs and allegations of the parties were heard at a meeting of a majority of all the members of said committee, and a majority of these present concurring in the following award: The said Arbitration Committee of the New York Produce Exchange do hereby award, order, and decide that after careful examination of the evidence Messrs. Libby & Clark have been justified in refusing the residuum so far tendered them by Messrs. Sone & Fleming and rejected on the contracts dated March 8, 1873, and that Messrs. Libby & Clark have not established any claim against Messrs. Sone & Fleming for damage upon the same, and that Messrs. Libby & Clark must pay the cost of first session $25, and balance of $175, each must pay $87 50.

In witness whereof, we, members of the said Arbitration Committee concurring therein, have hereunto subscribed our names this 2d day of January, 1874.

FREDERICK SHERWOOD,
A. S. JEWELL,
SAMUEL DALLY,
EDWARD HINCKEN,
WM. BLANCHARD.

Signed in presence of
WM. E. FLETCHER.

NEW YORK, November 12, 1873.

Arbitration Committee meeting held this day at 3 P. M. Present: Messrs. Sherwood, Hincken, Blanchard, Jewell and Dally. Mr. Sherwood in the chair.

The chairman stated that it was a matter in dispute with reference to a cargo of residuum, per bark "Norwood," from Philadelphia to Liverpool, between Messrs. Sinclaire & Marvin and Messrs. Warden, Frew & Co.

The plaintiffs stated that in January last they bought a cargo of residuum from the defendants at Philadelphia and shipped it to Liverpool, and it proved satisfactory. We received an order to buy another cargo of the same kind, and gave the order to the same broker to purchase it, which he did from the defendants (as per contract put in evidence, dated May 24, 1873), it to be free from coke and water, at the rate of 10 cents per each seven and one-half pounds. We sold the cargo on the other side at the rate of £9 per ton, cost, freight and insurance, and upon its arrival it was rejected, as it was not residuum, and not at all like the previous cargo, and we have received a certificate of its analysis, proving that it is not residuum. We landed the cargo and put it up at public auction, but failed to get a bid on it, and afterwards sold it for £8 5s, which was a very good price for it, but making a very large loss to us of £953 16s 10d. We have asked the defendants to pay us the damages, and they said they would look into it, and we have finally brought it here for settlement. The authorities on the other side ordered its removal as dangerous, which proved it was not residuum; and I will prove by witnesses what residuum is, and will now call Mr. Libby, of Libby & Clark.

Mr. Libby stated that they dealt very largely in residuum, and considered himself competent to judge what residuum is, and would say that it is the residue of crude oil after the heavy oil is removed, by distillation; and it should not contain more than from two and a half to four per cent volatile matter, and from 16 to 21 degrees gravity, free from naphtha, and thick like tar. Merchantable residuum should stand flash point at 300, and would not consider it pure residuum if it would burn in average atmosphere. I would not call this sample shown merchantable residuum, but it may be called residuum. It should vary about four degrees, from 17 to 21. We have found it as low as 14 and as high as 25. We buy it subject to inspection, and upon inspection, if it is not satisfactory, we reject it; and previous to May last the inspection required was to be free from coke and water.

Mr. Bunker, the broker, testified that he bought the cargo in January last, and also this cargo, which was to be the same as the previous one, and free from coke and water, and if it was merchantable residuum it would not flash at average temperature. We bought it subject to inspection, and nothing was said about the gravity of it. It is subject to buyer's inspection, and if he took it without inspection it would be at his own risk.

Mr. Bowring, of the firm of Bowring & Archibald, stated that they bought a cargo of residuum from the same parties about the same time, and contract was similar to this one, it to be free from coke and water and foreign matter, and upon its arrival at the other side it was rejected, as it contained spirits and other matter which it should not. We were not allowed to land it, and if it had been ordinary residuum we would not have been prevented. By examination here it contained 20 per cent of naphtha and at London 15 per cent. We had a sample taken of it, but did not inspect it, relying on the good name of Messrs. Warden, Frew & Co.

Mr. Lawrence, broker, testified: I have made a great many contracts for residuum during the past eight years, and it is the refuse from crude oil after first distillation, and should be thick like tar. I would not call the sample shown a merchantable residuum.

Mr. Wheelock, inspector, testified: We received instructions from Messrs. Sinclaire & Marvin to inspect this residuum, and it was not to exceed 950, and to be free from coke and water. I made a test and found it from 880 to 885, and telegraphed to Sinclaire & Marvin, and they replied by letter to ship it. I should say this residuum contained naphtha, and would not accept it now, as at that time we did not know anything about gravity, and had no experience at the time, having only inspected one cargo previously. I supposed it was a merchantable article at the time, but do not now think it is.

The defendants replied that they filled this contract as they fill all others, and it was made up from three different refineries, and it was not all inspected on our wharf, yet they report it was all the same quality, which seems very strange; for if one refinery made a poor article, it is not at all likely that they would all make the same at the same time. Our work is not done in the dark, and our refinery is open for inspection at all hours, and all we put in our tanks is water to raise the contents up, which operation is done in the presence of every one there and done every day. We would not be likely to put naphtha in it, when naphtha is worth 25 per cent more. We sold this in New York for a good article, and they came down and inspected it, and put it in their own vessel, and it was all done under their own supervision, and we have filled the order according to contract.

The plaintiffs further stated that they did not think the defendants put anything in it knowingly, but it is in there, and how it came there we do not know.

After the presentation of the above evidence, the Committee made the following award: That, according to the evidence presented, Messrs. Sinclaire & Marvin have failed to establish any claim against Messrs. Warden, Frew & Co., and they must pay the cost of this arbitration, $25.

In witness whereof, we, members of the said Arbitration Committee concurring therein, have hereunto subscribed our names this 18th day of November, 1874.

FREDERICK SHERWOOD,
A. S. JEWELL,
EDWARD HINCKEN,
SAMUEL DALLY,
WM. BLANCHARD.

Signed in presence of WM. E. FLETCHER.

IMPORTATIONS, 1873.

The following is a complete list of the imports of Oils, Paints and Drugs entered at the ports of New York and Boston. (In cases where weight cannot be correctly stated, the number of packages alone be given):

NEW YORK.

Item	Pkgs.	Lbs.
...te Root, dks.	27	5,129
" Leas, pkgs.	8	1046
" Herbs, bale.	1	115
" Rad, bales	2	482
...d, ck	1	20
" Arsenic, pk.	174	137101
" rdde, cks.	7	11,285
" Boracic, tcs	571	633,963
" Benzoic, cs	2	...
" cs	10	293
" Gic, pkgs.	810	88,458
" Carbolic, cs	621	...
" cs	237	35,978
" Fluoric, cs	4	...
" Lactic chnl, cs.	11	450
...nl, cs	6	281
" Not specified, cs	9	...
" Oxalic, dks.	812	534,763
" Pyrogallic, cs	8	125
" Hc, pkgs	19	...
" cs	53	11,206
" Tartaric, cs	37	7,552
" Phosphoric, cs	3	336
" Rosolic, pkgs	46	10,533
" Hc, cs	1	56
Adhesive Plaster, cs	7	...

Item	Pkgs.	Lbs.
Agaric, pkgs	6	827
Albumen, pkgs	444	107,951
" pkgs	402	...
" Black, pkgs.	111	11,200
" bgs	50	111,928
" Bld, bgs	508	...
" pkgs	121	...
" Egg, cs	121	24,865
" cs	44	...
" Vine Chlorate, cs	1	...
Alzarine, cks	220	89,799
" cks	38	...
Alkali, cks	1192	1,456,703
" Refined, pkgs	161	204,590
Alkaline Blue, cs	9	547
" Colors, cs	12	770
Alkamet, bls	2	822
" Root, bales	71	15,579
Aloes, pkg	1029	77,991
" cs	207	66,985
Althea Root, cks	20	...
" ck	1	b21
Aluminous Cake, pkgs	1435	1, 87(.3
Alum, ck	352	225,017
" cake, cs	115	72,892
" brnl, ck	181	...

Item	Pkgs.	Lbs.
Ambergris, pkgs	3	oz 62
" Sal, ck	446	459,859
Ammonia, ck	2	...
" Carb, cs	1080	605194
" Muri, cks.	237	168,732
" Mte of, bbls.	6	2271
" Spte, cks.	620	700192
Analina, cks	27	22,988
Matto, pkg	1789	106,373
" pkgs	34	...
" Seed, bbls	24½	...
" bbls	10	46944
Angostura Bitters, cs	300	gal 675
Aniline, cs	16	1552
" pkgs	14	...
" Mte of, cks.	24	16,159
" Bon, pkg	1	...
" Cake, cs	3	...
" Crude, cks	10	3,774
" pkgs	21	...
" Krs, pkgs	729	59,982
" Dye, pkgs	358	83,995
" sad, bags	742	27,672
" bags	323	...
" Star, cs	110	13,998
Antichlor, cks.	53	...

Article			
Antimony, Sulph., ck.	1	250	62
" cks.	50	30000	1
" Crocus, kegs.	10	1250	11
" Regulus, cks.	182	93100	1
" cks.	33	....	
Archill, pck.	64	17,062	23,787
" pkgs.	53	....	11,979
" Rad, pkgs.	40	18089	6
" Extract cks.	15	....	135
" cks.	182	106,379	476
" Liquor, cks.	4	....	15
" cks.	569	145625	500
" Weed, bgs.	38	....	1,096
" pkg.			725
Argols, pkgs.	2550	2,152,307	230
" Crude, cks.	2907	1,934,229	82
" part Ref'd, cks.	1389	1,249,431	24
Arnica Flowers, bales.	156	30,407	5
" bales.	67	....	37
Arsenic, pkgs.	1260	608,167	....
" Crude cks.	66	25,977	634
" Phosphorus cs.	1	....	22
Arrowroot, pkg.	1229	102,859	1
" bbls.	70½	....	5166
Assafœtida, cs.	513	107,092	1042
" cs.	70	....	5
Alum, pkg.	475	345,947	86
" crude.		1008000	4350
Aurine, pkgs.	254	39,513	10
Balsam, cs.	7	875	15
" Peruvian, ck.	1	....	4
" cs.	77	7940	180
" Capivi, pkgs.	411	46,024	100
" Capaiva, cs.	111	....	124
" Toin, cs.	102	11221	42
Bark, bales.	542	52957	10

Article			
Bark, bales.	250	....	6528
" Buckthorn, bale.	30000	200	110
" Mezereon, bale.	1250	2588	1300
" Bittersweet, bale.	93100	50	13,027
" Peruvian, pck.	17,062	....	452
" pck.	18089	866	690
" Calisaya, cer.	106,379	600	4,700
" cer.	145625	....	204,516
" Quilla, bales.		94,000	8 680
" Columbia, bales.	2,152,307	....	112
" Quinine, pkg.	1,934,229	41800	4400
" bales.	1,249,431	....	....
Barwood.	30,407	76739	4496
Barytes, cks.	608,167	29,849	2000
" cks.	25,977	....	400
" chlor, cs.	....	2275	2,000
" Sulphate of, cks.	102,859	503	3,696
" cs.	107,092	19600	....
Barium, Chlorate of. bbls.	....	50000	22074
Berbery, Ex. of, pkg.	18	....	27,013,448
Bergamot Paper, cs.	1	200	5100
Belladonna, bales.	9	2588	1608
Belladonna Leaves, bas.	75	50	918
Belladonna Root, bales.	3	....	692
Bel ina Heros, bales.	3	866	....
" pkgs.	25	600	12716
Berries, Yellow, bgs.	751	....	....
Binoxalate, cks.	31	94,000	4,095
Bismark Brown, pkg.	1	....	596
" cs.	2	41800	299,000
" kgs.	2	....	....
" cs.	17	76739	400
Bitter Sweet Root, bales.	13	29,849	....
Bitter Sweet Stems, bales.	4	....	2369
Bitt Wd.		2275	64,712,400
Black Mr, pkgs.	30	503	
Black Paste, ck.	1	19600	
Blanc Fix.	33	50000	
Edh, pkgs.	30,248	938,883	
Blue Galls, bgs.	20	20261	
Ble Verditer, kgs.	14		
Bombay, Ext's, bbls.	2	2,819,686	
Bone ash, pkgs.	2	769,146	
Bone Bik, cks.	25	500	
Bone cks.	16	53,101	
Borax, tcs.	4	....	
" cs.	22		
" Refined, cs.	2	1638	
Braziletto Wood.		564	
" pieces.	1311	66,392	
Bremen Black, cks.	3	....	
" Blue, cs.	24	10096	
" pkgs.	31	2230	
Brimstone,			

NEW YORK IMPORTATIONS CONTINUED.

Item	Pkgs.	Lbs.
Bronze Green, pkgs.	10	1338
" cs.	50	498
Leaves, lbs.	179	57,353
" lbs.	49	...
Burgundy Pitch, tubs.	350	39,430
Buttonlac, chts.	148	35,717
jhs.	70	7,511
ck.	1	462
cs.	39	...
pkgs.	3	200
fr, cs.	4084	397,643
cs.	5419	...
Crude, bxs.	394	69,747
e.	...	60
lli	15	64
cs.	7,031	71
Canary Seed, pkg.	1470	...
pkg.	43	8 28
cs, pkg.	1091	...
Seed, pkg.	4595	30
jhs.	...	34
e of Iron, cks.	1	30
cs, cs.	128	...
Carmine, jhs.	9	...
pkgs.	4	160
Clors, ck.	1	110
Dry, pkg.	15	1254
Dry, jhs.	50	...
of Indigo, pkgs.	11	53
Lead, cks.	5	42
Re, cs.	25	43
cs.	2	...
Scarlet, keg.	1	92

Item	Pkgs.	Lbs.
Gia Fistula, bale.	38	20
" Buds, cs.	50	3333
fri, cs.	6	...
" cs.	...	235
Dry Seed, pgs.	4	511
Wu Blue, ck.	1	...
thry Herbs, lbs.	3	40
Mr, b.les.	15	2210
Chalk	...	10546853
afk, kk	...	22,467,000
" French, pkgs.	155	39,268
Gmomile Flowers, cks	273	...
" cks	128	...
Chille, bales	13	80
We Blue, jhs.	15	64
M, Ha.	2	31
Ho. Hydrat, cs.	255	16,269
Wr. Kalium, ks.	8	58
Chloride of Wc. cs.	20	92
" e.	11	...
" cs.	4	34
Chrome Yellow, pkg.	48	30
" cs	4	...
" Green, pkg.	8	07
" Orange, pkg.	1	03
Wl Es, ks.	25	20,330
Chufus Rd, jhs.	50	50
Clay China, jhs.	6720	4,881,166
" jhs.	2296	2, 88 84
" Fire, cs.	121	176,000
Glt, cs.	33	33
Ws Indicus, bales.	441	28

Item	Pkgs.	Lbs.
Cochineal, pkgs.	5092	899744
" pkgs.	2449	...
" Teneriffe, cks.	5	5759
" cf, ck.	1	...
Win Lake, cs.	1	220
" cs.	2	...
Ga Bar, cs.	9	1508
Colchicum Seed, cks.	4	1050
" Rt, pkg.	6	1166
" Wine of, cks.	4	400
Colchi Rt lbs.	8	1,908
Coi tinc, pkgs	1734	154,308
" pkgs.	284	...
Col rjh, cs.	21	4966
" pkgs.	8	...
Colombo Rt. bales.	259	22800
Coi e, pkgs.	1523	337,403
" jhs.	264	...
" Dry, cks.	9	4612
" e.	3	...
" lbs', cs.	12	2515
" tbe e.	4	400
Irg, cks.	5	261
st Herbs, bale.	1	100
Kfn, ble.	1	110
nr Seed, bl cs.	1103	188.313
Cosmetics, e.	2	...
Gr, Mate 6, cs.	3	596
Gs, cks.	79	76.078
Covnage, Gm cr, lg.	1527	25
nz White, pkgs.	6	1723319

Item		
Cremnitz White, cs	2	500
" Yllw, pkgs	2	....
Crocus Powder, cks	10	5000
Crystals, part refined, cks	6	5,840
" Yllw, cs	1	100
Cubebs, bgs	648	219,543
" Gr, pkgs	335	222087
Cummin Seed, bgs	453	56,873
Cundurango, bales	7	186
Cutch. pkgs	14,536	2366191
" bxs	4250	....
Cuttlefish Bone, pkg	22	....
" pkg	359	19,062
Dandelion bls, bales	4	800
Dandelion Root, dhes	18	3733
Dextrine, cks	84	85,752
Digitalis Leaves, dhcs	8	1410
" Herb, dhes	2	357
Disinfecting Powder, cks	2	224
Divi Divi, bgs	7598	8527,008
Dog Grass, dhe	1	....
Dragon's Blood, cs	29	3,345
Dried Roots, dhes	13	3,900
Drop blk, pkgs	804	75,827
" cks	265	....
Drugs, etc., pkgs	37	....
" Crude, cks	35	8131
" cs	17	....
Dry Herbs, dhes	106	....
" pkgs	6	720
Dutch Ochre, cks	315	187,793
Dutch Pink. pkgs	31	3232
Dyestuff, pkgs	6	....
" pkrs	39	6,910
" hhd, cks	10	....

	Item	
34	Dye Flowers pkgs	....
2	" bns	220
277	Dyewood, pkgs	....
17	Dyes, cs	1275
107	Earth Silicate, cks	....
95	" cks	200
1	Elder flrs, bale	100
15	Emerald Gen, cks	9,324
25	Emery Flour, cs	5454
27	Ergot, pkg	5,051
2	" cks	....
114	Ergot of Rye, pkg	16,807
20	Essence lia, cs	gross 120
33	Lavender, cs	2700
12	" pkgs	....
162	Red , cs	6655
5	Thyme, pkgs	....
16	" cs	1218
8	Rosemary, cs	225
4	" pkgs	....
1	Rose, cs	113
2	" cs	....
19	Lemon. cs	....
99	" cs	2188
2	Pepsin, cs	500
3	" cs	....
27	Orange, cs	677
114	not specified, cs	....
45	" cs	2653
1	Geranium, cs	22
14	Bergamot, cs	411
20	" cs	230
4	Romarin, cs	oz 35
1	Neroly, pkgs	450
10	Origanum, cs	....

Item		
Essence Sage, cs	1	95
" Sabine, can	7	....
Ether, cs	5	219
" cs	3	....
" Formic, cks	10	10
" Butter	2	200
Eualyp'us dhs, cs	10	....
Extracts, cs	10	4510
Extract Barbary, bbls	75	33,157
" On, cks	15	oz 100
" Cuba, bbls	1	9678
" of lM, ck	25	156
" flM, bbls	2	60
" Belladonna, cs	1	....
" Mlr, cs	1	20823
Malt, cs	50	2120
Persian Berries	16	4441
Fustic, bbls	10	56,017
" Tannic, bbls	266	100
Fennel Seed, dhes	1	....
Fern Root, bale	2	187
Fish, Sound Raw, bx	3	20850
" dhrs, bbls	115	3441
Filling up, pkgs	6	700
Florentine dl, cks	3	....
Fluer Spar, cks	17	251,250
Fuller's Earth. bgs	1628	9,895,187
" M, pks	70,637	....
Foenegreck Seed, bgs	60	....
" bgs	278	59,883
Fox Gd, dhe	5	1130
" Wt Black, cks	122	66756
" pls	44	....

NEW YORK IMPORTATIONS CONTINUED.

Article	Lbs.	Pkgs.
French ... ck	1	217
Galangul Root, bales	100	333
Galbannne, cs	3	~0
Galls, bgs	127	31,577
Galls, ...	40	....
Gambier ...	7,521	....
" Bas	2,079	980,021
Gamboge, pkgs	177	18,705
" bxs	25	....
Garancine, pkgs	97	....
" cks	5,762	7,820,198
Man Root	508	126,883
" ...	5	..
" ...	278	..
" ...	139	11,992
Ginger, bxs	2,072	67,301
" ...	215	....
" Wte, pkgs	339	484
" Jamaica, bbls	145	15,747
Glue, pkgs	2,016	860,668
" substitute, cks	155	4000
" ... cks	4,584	1,789,547
" pkgs	265	....
Glycerine, pkgs	2,475	838,638
" cs	255	....
" ... cks	10	12,786
Gogo, bales	20	4,104
Gold Size, pkgs	42	959
" Paint, Bronze Powder, cs	3	....
Grape Sugar, cs	29,368	2,714,056
Grease, cks	42	13,400

Article	Lbs.	Pkgs.
Milori, cs	...	929
Guaza, bales	36	3,716
Guinea Grains, bgs	...	3,400
Gum, pkgs	531	77,766
" pkgs	405	....
" ..c, cs	7	1,145
" ...	25	13,641
" rabic, pkgs	5,721	2,638,645
" pkgs	306	....
" siftings, pkgs	201	1 4,880
" Assafoetida, cs	20	7,97
" Bark, bags	8	2,599
" Benjamin, cs	6	1,458
" Benzoin, cs	3	170
" Copal, cs	1,039	....
" bbls	1,431	295,542
" Damar, bxs	17	....
" pkgs	292	32,252
" Elems, cs	2	407
" Euphorbium, cers	2	460
" ... bales	181	856?
Guiac, pkgs	50	....
" pkgs	21	3,258
Guiacuim, bbls	21	....
Kino, cs	13	2,744
Kowrie, pkgs	10,085	2,652,557
" cs	3,594	....
" ..m, cs	138	12,110
Mastic, cs	5	493
" bbls	2	....
Mogadore, bbls	20	8,158

Article	Lbs.	Pkgs.
Gum, Myrrh, bales	108	15,160
" Sandrac, cks	52	19,212
" Senari, cs	13	4,044
" Senegal, bgs	213	44,671
" Shellac, chts	50	10,567
" cs	10	....
" Substitute, cks	39	40,775
" ..c, cer	987	4044
" bgs	200	....
" Siftings cks	30	19,814
" Tragacanth, cs	216	4,951
" Yellow, cs	10	3,145
" ... cks	3,706	7,190,385
Hartshorn Shavings, bales	1	2,970
Hellebore Root, bales	11	2,400
Hemp ..d, pkgs	1,638	313,461
Herbs, .c, bales	7	1,300
" Centaur, bales	3	625
Horehound, bales	5	997
Horehound Leaves, bales	3	525
Humboldt, cs	1	100
Hyascuamus, cs	15	3,930
Ind ..s, bales	8	1,168
Ian Red, firkins	10	....
" cks	4	5,005
Indigo, pkgs	4,355	681,763
" pkgs	449	....
" Acid Extract, cks	28	18,528
" Extract, cks	222	106,920
" pkgs	39	....
" Paste, cks	285	135,029

Left table

Pkgs	Article	Value
35	dingo, Paste, pkgs	....
2	Indigotine, pkgs	5676
1	Indulin, cs	200
708	Insect flr, pkgs	63,603
234	Iodine, pkgs	21,519
81	" Crude, pk	15,936
10	" " cs	....
11	" Green, cs	920
92	Epc, pkgs	8,544
136	Ipecacuanha, cer	17,504
878	Isinglass, pkgs	107,580
220	" pgs	....
33	Ivory Black, pkgs	6,319
18	" pkgs	....
95	Jalap, cs	17,902
22	" cks	....
1	Japanese Brown, ck	....
591	iper Berries, bales	89,147
680	" bgs	165
1	" Juice, ck	....
500	Kaiaet, bags	110,236
10	Kaoline, cks	12,568
33	Kremm, White, cs	6,627
20	" pkgs	1,995
129	Kreosote, cs	22,612
107	Lac Dye, pkgs	....
34	" chts	7,501
7	Lake's, pkgs	3,130
2	Lake Wd, cks	....
28	" bk, cs	761
25	Lapis Calaminous, kegs	6,034
20	Laque tin, cs	8,557
41	Laurel Berries bales	11,736
213	" Leaves, bales	65,120
3	Lavender Essential, cs	....

Right table

Qty	Article	Value	Pkgs	Value
152	Lavender Flowers, bales	44,057	1	100
414	Lead, Nitrate of, pkgs	348,017	2	264
1	" Black, cs	....	81	9,804
20	Leaves, bales	....	24	....
57	Lemon Juice, pipes	gals. 12,711	174	17,624
5½	" Peel, cs	....	1,041	124,023
20	" Crude, pk	4,743	5	....
387	" not spec, cs	45,923	15	750
71	" Juice, kegs	15,450	2	4,257
5,864	" Paste, cs	1,492,509	28	21,163
92	" kgs	7,999,675	133	....
57,207	" Root, pkgs	66,936	30	....
4,451	" pkgs	506,000	10	6,274
565	" Sticks, cs	2,600	34	1,383
....	Lima Wd	947,202	11	50
4	fie Muri, bbls	2,600	1	425
1,045	Cate of, cks	800	2	281
4	" Carbonate of, cks	6,114	1	....
100	" Chte of, pkgs	683,950 bush	1	549
14	" Juice, phns	17,764	5	1,098
10	" fie of, cks	5,000	8	27,412
1	" Vma, ck	48,452,343	179	6,532
597,929	Lad, bgs	36,243	33	18,149
74	Litharge, cks	6,642	85	....
10	Litharge Flake, cs	1,833	7	45,305
37	Gld File, cks	836,899	244	....
-	Logwood, cs	....	80	13,445
110	" Extract, cks	167,241	86	6,045
27	Lycopodium, cs	72,622	54	15,000
16	" Seed, cks	....	150	1,000
575	Maddcr, cs	....	5	1,200
155	" cks	....	15	549
248	" Extract, kegs	....	7	....
11	Manganese, cks	....	6	....
90	" pkgs	....	20	....

NEW YORK IMPORTATIONS CONTINUED.

Item	Lbs.	Pkgs.
Mineral, Crude, cks	8	....
Mineral [Ign], cks	3	67200
" Earth, cks	120	11774
" [Wite], cks	16	oz 360
[Fine] Sulph, cs	2	650
Munich Lakes, cs	2	
Munjeet, bales	10	....
Musk, boxes	12	273
" Crude, cs	2	oz 554
[Mdl cSd, pks]	5	593447
" "	2771	bush 1331
" "	240	
Myrabolane, bags	440	
Myrrh, cs	20	500
Myrtle Berries, cs	2	100
Naples Yellow, pkg	1	
Napthaline, yellow, keg	1	50
" brown, cs	17	1650
Natron Carb, cks	1	140
Nitro Benzoli, cs	47	7200
Nutgalls, bgs	807	163414
Nux [Mia], bgs	94	
" bgs	1691	117260
Oak Stain, bx	3	500
" [bd], cks	20	
" cks	2473	1788492
" [Rd], cks	11	11636
" [Kn], cks	25	13200
" [Kn], cks	11	
" It [inn, cs]	2	900

Item	Lbs.	Pkgs.
Ochre, Yellow, cks	150	13975
Oil Almonds cs	74	28028
" cs	21	
" sweet, cs	8	800
" bitter, cs	2	40
" Amber, cs	13	1191
" [fnd], cs	2	450
" Aniline, cks	283	129,578
" Animal, ck	39	92755
" Aniseed, cs	2	643
" pkgs	33	
" Star boxes	87	6398
" cs	20	1332
" Anise, German, cs	60	
" Aspic, cs	1	5
" Balsam Capivi, cs	1	
" Bergamot, pack	1105	10
" Cajeput, bbl	144	53210
" cs	34	
" Camphor, cs	20	729
" [nhs], cs	1	6
" Capivi, cs	5	
" cs	1	50
" Cassia, cs	1	
" Castor, cs	202	13393
" pkg	729	gal 5,238
" Caraway, cs	252	5455
" Ceder, cs	83	101
" [Komomile], cs	3	2
	1	

Item	Lbs.	Pkgs.
Oil, Cinnamon, cs	7	....
" cs	2	28
" Citronella, cs	222	9406
" pkg	....	oz 56886
" pks	421	850
Cloves, cs	17	300
Gal Tar, cs	2	
[Gal Tr], pkgs	3	
[chut]	712	gal 902788
" cks	163	
Codfish, cks	25	gal 1,450
Cod [Tar], cs	704	
" cks	454	galls 28419
Cologne, cs	1	
" cs	1	10
Colors, cs	5	
" cs	2	300
" [Gd], cks	6	
" [ton], cs	40	gal 256
Cummen, cs	3	1980
[Bfc, cb]	4	98
Essential, cs	128	7335
" box	42	
[Fgs], cs	1	112
Extracts, cs	1	
Fennel, cs	8	400
Fennel Seed, cs	1	10
Fish	20	40,000
[Kh], cks	11	gals. 937
[Fsil], cs	4	gal 32
Gallipoli, cks	7	gal 533

Oil, Geranium, cs	12	271	Oil, Palm nut, cks	19	11000	Oil, Valerian, cs	1	20
" Ginger Grass, cs	1	25	" Palmarosa, cks	1		" [illegible], cs	1	51
" Haarlem, cs	199	....	" Patchouly, cs	2	28	" Whale, cks	1010	237466
" " cs	10	108	" "	5		" [illegible], cks	202	....
" Jasmine, cs	2	191	" Peppermint	1	4	" [illegible], cs	1	4
" Juniper, cs	48	4209	" Pimento, cs	1	20	" Ylang Ylang, cs	1	1-3
" " pkgs	7	....	" [illegible], cks	5	1662	Olives, cs	110	
" Juniper Berries, cs	11	812	" " pkgs	6	....	Opium, cs	1124	167405
" Juniper Wood, cs	21	2304	" Portugal, cs	10	...	Orange Lead, cs	886	440364
" Kagn Putch, cs	4	...	" Rapeseed, cs	7	1327	" Mineral, pkgs	542	264746
" [illegible], cs	70		" Red [illegible], cs	32	1768	" [illegible], pkg	113	
" " do	171	15176	" Rhodium	3	120	" Peel, bales	135	24738
" Laurel, pkg	22	2,180	" Rhodium, pkgs	1	....	" " cs	12	
" Lemon, pack	1925	51627	" Rock, cs	1	50	Oris Root, cas	123	50209
" " "	136	....	" Rose, cs	17	oz. 8182	Orseille, pkgs	16	1833
" Lemongrass, cs	8	oz 4682	" " pk	6	....	" cks	15	
" " cs	178		" Rosemary, cs	49	4664	Otto Rose, cs	29	344
" Linseed raw, pkgs	70	gal 14834	" Rye, pkg	1	15	" " cs	4	
" Linseed boiled, pkgs	3	gals 40	" Salad, bx	55	....	Oxalate [illegible], cs	1	oz 10
" Mace, cs	1	20	" " cs	5	gall 7	Oxide of Iron, pkg	142	27170
" Mirbane, cs	271	28850	" Sandelwood, cs	28	1644	" pkgs	122	
" " cs	37		" Sassafras, cs	2	158	" in oil, pkgs	54	8360
" Moon, ck	1	360	" Savine, cs	3	206	" [illegible], cs	3	230
" Neroli, cs	3	14 1-5	" Seal, cks	153	galls 20157	Oxydizing Paste, kgs	108	13315
" Nutmeg, cs	2	64	" Sesame, cs	168	25,951	" cks	12	
" Olive, pkg	16094	galls 144,282	" " bbls	28	gal 1404	Paints, cks	62	31223
" " "	19975	....	" " cs	50		" pkg	105	
" " cs	233	3362	" Sod, cks	91	gal 20208	" Brown, pkgs	25	
" Orange, pkgs	313	8008	" " cs	47	....	" [illegible], cks	10	5664
" " pkgs	16	....	" Sperm, cks	82	gal 16886	" Imperial [illegible], cks	10	6677
" " [illegible], pack	22	487	" Sweet, pkgs	4	....	" Patent, pkgs	47	20117
" " sweet, pkg	31	914	" Tanners, cs	11	2700	" Red cs	5	
" Otto Roses, cs	2	46	" Thyme, cs	128	10160	" " kegs	40	4100
" Palm, cks	2		" " cs	5		" Tubes, cs	5	
" " cs	33	gal 6350	" Valerian, cs	1	....	" White, pkgs	300	54019

NEW YORK·IMPORTATIONS CONTINUED.

	Lbs.	Pkgs.
Paints, Yellow, pkg..	1	299b.
" dks..	78	
Pareira Brava, pkgs	8	535
Paris White, cks	3366	H997?
" 6d, pkgs	4	668
" bbls	7	
Mnt Blue, cs..	2	
" Barley, cks	10	
" Dryer, cks	467	269431
" cks, cs	3	
Patros cds bags	28	3900
Peppermint Herbs	12	1300
Pepsin, cs..	1	
Persian, Red, pkg	104	3507
" cks	17	
" Berries, bgs	24	
Persian Berries, bags	5	800
Phosphorus, pkg	602	6844
Picric Crystals, cks	2	533
Pigment, cks	5	1,120
" cs	1	
" Brown cks	13	7637
" Green, cks	18	7541
" Orange, pkg..	7	
" cks	23	12308
Plumbago, bbls	20,173	9,123,557
" cs	69	1100
Ponceau, cs..	16	
Poppy Heads, cs	28	4,006
" bgs	5	
Poppy Seed, pkg..	463	87,720

	L's.	Pkgs.
Poppy Seed Flour, bales	40	2594
Omli Leaves, cdes..	2	24
Portland I cdt, cks	1977	
" bbls	300	107000
Potash, hto, pkg	4,701	672,994
" Hi, cks	1274	1,000,176
" Binoxalate, cks	15	3000
" Iodide of, cs	25	1450
" Mri, bgs	38,346	8,541,670
" Premangrate, cs..	1	12
" Prussiate of, cks	222	103552
" Prussiate W, cks	211	118,589
" Refined, cks..	32	50636
" Sulphate of, bgs	1,128	
" pkgs	1,347	314272
Printing Black, cks..	4	
Pumice Sae, cks	180	1267
Pple Ros, bxs	1	28
Wood W, bbls	10	8,087
Prussian Blue, pkg	48	7407
Pyrethri Flowers, bales	15	2,229
Quaker Kan, pegs.	5	711
Quercitron, bbls	10	
Quicksilver, bdes	600	
Quinine, cs	375	28750
" cs	319	7197
Qivinoy Seed, bales	29	
" cs	1	117
Radtaraxer Roots, bales	4	800
Rape Seed, cks	657	141386
Redwood..		51900

	Lbs.	Pkgs.
Red Lac, ck	1	
" Lead, pgs	713	84,756
" pkgs	30	217
" ldr, ck	1	94960
" Saunders Wood, pkgs		1921
Rhatany Root, bales	19	53,826
Rhubarb, pkg	342	
" bxs	132	1109
Rosemary laves, lb	5	263
Rose Leaves, pkgs..	3	
" Leaves cs	2	
" Pik, cks	295	79,302
" cks..	65	
" water, cs	56	
Root Flour, bxs	25	6000
Roots pkgs	4	723
Roth cd, cs..	1	
Rotten Stone, cks	22	100000
Ssflower, bales	153	38,916
" Extract of, pkg..	38	55,556
" cs	245	
" cs	5	gall 125
Saffron, pkg..	39	3378
Sage Leaves, bales	215	26301
Sago, cks..	245	68,505
" cks	265	
" brs bags	975	
Salacetos, cks..	1	224
Salts, Aniline, pkg	59	16,695
" cs	4	
" Glauber, ck	1	864

	Item	
177	Salts, Black, pkg.........	
25	" " kegs..........	
76	" Cal, pkg.....	
10	" Crude, pkg.....	
90	" Epson, bls..........	
12439	" Manuring, pkgs......	
501	" " bgs......	
4	" " cs.......	
492	" Washed, bgs......	
20	" " pkgs.....	
6	Salt Cake, cks............	
28712	Saltpetre, pkg....	
23772	" Crude, bgs.....	
11	" " pks	
200	" Refined bbls.	
63012	Sal Soda, pkg.......	
2	Sandlewood, cs.....	
5	Santonine, cs	
468	Sapan Wood........	
10	Sapan wd, pieces...	
3004	Sapan wd, extract, cks.....	
603	Sarsaparilla, lbs.......	
45	" pkgs.......	
3	Satin White, cks.....	
1	Scammony, cs........	
6	" cs	
12	Secale Root, pkgs......	
1	Sedlitz Powdered, cs...	
58	Sclery Seed, bales......	
50	Senna Leaves, bales.....	
631	" bles.....	
5	Senna, pkg.....	
4354	" burnt, cks.....	
1655	Shellac, cs.....	
	" pkgs.....	

	Item	
45,599	Sienna Earth, cks. ...	240
38071	Size, cks....	2
3,492	Smalts, cks.....	63
34,362	" cks.....	24
2138949	Soap, Castile, pkg	20
	" bbls....	61304
180	" hxs....	1674
	" Root, sks...	38
2106	Soda cks. ...	6
13157	" Ash, pkg	5408
5603158	" cate of, cks..	51695
2359680	" Bi-Carb. pkg....	60
22400	" Binarseniate, cks...	88753
	" Bisulphate, cks...	18
19411187	" Caustic, dms....	4
	" ite of, ck	27488
73	" Hypo. Pulvia. cks	1
250	" " cks...	70
509881	" Hypo. Sulph. cks	60
	" cs...	135
4576	" Hyposulpate cks...	1
396803	" Nitrate of, bgs	956
	" " bgs...	63942
33615	" Silicate of, cks...	29673
	" " pkg	39
50	" Stanate of, cks...	29
1045	Sponges, cs...	17
3360	Square Black, cks...	2
115	St. John's, bad, cks...	4
18637	" " bbls...	324
172408	Sterine Candles, pkgs...	105
2400	Sticklac, cs...	120
756460	" cs...	54

	Item		
156670	Styrax. bbls....	9	1641
	Styrax Liquid, cks	4	1322
66523	" " bbls....	1	....
	Strontia Nitrate, pkgs,...	2791	827,600
2210	" " cks.	10	1500
2014518	" Mineral Crude....	5	5707
8556	Suga. of Milk, cs...	19	280
	Sulphur, bbls....	1	6827
	Sulphur Lac, bbls...	25	
2256952	Sumac, bgs.....	41311	5481096
77338526	" bgs ...	10,217	189164
77189	" Extract, cks...	508	100,000
997	Superphosphate, bbls	400	981
9752	" bc. bgs...	981	137452
3665	" cks...	465	
17402666	Tapioca, bgs....	421	68050
	" bgs....	486	968
33,156	Tarazica Rad, bales...	4	12,200
	Tartar ft Refined, cks...	13	19830
150,346	" bags...	20	61,467
	" cks...	16	1918
387189	" Crude, cks...	114	11341
59925094	" Emetic, pkg	5	
25994	" Refined, cs...	10	283
	Taxicum Root, bgs...	3	9623055
13638	" cs...	1	....
	Tra Alba, pkg	2748	2 119
900	" cks...	1699	
51736	Tra Umber, cks...	50	3341
160	" cks...	31	92
	Thyme, tbs...	17	12952
7111	Tra Flowers, pkg	1	730
21,805	Tin cks, pkgs...	26	1
	Tin Muriate, ck...	1	
	Tongua Butter	1	

BOSTON IMPORTATIONS.

	Lbs.	Pkgs.		Lbs.	Pkgs.
Tripoli, cks	29	...	White Lead, in oil, bbls	350	53404
Tuduline, cks	2	547	" pkgs	78	...
Wc, pkg	2,952	390,528	" not specified, pkg	13942	4554615
" bgs	1,000	...	" cs	862	...
Turpentine, bls	20	gal 235	Witherl e, cks	160	80000
" Venice, stand	253	34,052	Wo d, cks	38	56002
Tuscan Red. cks	55	6347	Wc cl Lake, cks	50	17738
" cs	20	...	" pkgs	33	...
Ultramarine, pkg	3436	1021029	Worm Seed, bales	232	46929
" Wa, cs	11	4169	" Ues	107	...
Umber, cks	140	76642	Zaffer, cks	18	1985
" Burnt, cks	24	8,530	Zinc, Oxide, bbls	13299	2931509
Urva Ursi Ws, bales	38	8073	" Dust cks	10	10800
Wn, Red, : cs	7	1784	" Drop Skimmings, pkgs	369	...
Wan Root, bales	63	81389	" Lactate, pkg	1	...
Van Dyke Brown, ck	31	22777	" Sulphate of, cks	47	31458
" cks	23	...	" White, pks	60	6600

	Lbs.	Pkgs.
Vanilla, bx	1	...
Varnish, cs	482	gal 15554
" cs	85	...
Vegetable Wax, boxes	157	3193930
Wan Red, bbls	9992	...
Wt Brown, cks	1	634
Verbasa Fels, cks	7	...
Verdegris, cks	58	6?9?6
Verdegris, cs	5	pkgs
Win, bxs	26	...
Vermilion, pkg	482	81085
Wash Blue, cs	260	...
" cs	28	6462
Wg Powder, cs	2	500
Wer cks, cs	119	...
" bls	22	4606
White Lead, dry, bbls	359	227744

BOSTON.

	Lbs.	Pkgs.		Lbs.	Pkgs.
Ad fc, kgs	65	7280	Wnia Sal, cks	5	...
" Carbolic, cs	4	...	" Sph, cks	10	13070
" Oxalic, cks	212	175304	" cks	24	...
" Picric, cs	8	1512	Annatto. cks	4	...
" Wfc, cs	3	300	" Liquor, cs	1	...
" Tartaric. cks	2	1120	" Paste, cs	1	666
Ave Wter, cs	1	100	Aniline pkg	1	...
Wn, Egg, cs	29	6530	" Marrow, ck	1	...
" Wd, cks	11	2405	" Soda, ck	1	710
" pkgs	12	...	Argols, pkgs	7133	587249

Item			
Argols, Crude, cks........	201	196210	68
Arsenic, cks........	204	391822	208
" bbls	13	3,323	250
Ashes, Pearl, pkg........	58		35
Ashes, Refined, cks........	133	103942	4582
Alkali, cks........	60	72010	33
" tins Cake, pkgs........	549	393585	
Antimony, cs........	125		5
Berries, Yellow, bgs........	856	311667	1
" bgs........	201		3
Binoxolate, cks........	101	22736	1
Bith, cks........	10816	8460798	2
Blue Galls, bgs........	41	3862	22
Brazil wood........		254000	3
Brimstone	10,	61307	213
Buchu Leaves, bales........	10		115
Camphor	167		
" bxs		2000	29953
Camwood		57227	1018
Canary Seed, cks........	50	8975	15
" bgs........	50		237
Carmine, cs	1		113
Celestial Blue, kgs........	35	4522	10
Silk, cks........	2	595	2900
Conal Salts, cs	8	4480	4
Cochineal, bgs........	266	50125	135
Colocthar, kgs........	170	22,975	75
Colocynth, cs........	10	1470½	12
Colors, cks........	33		109
" dry, cs........	1	50	2429
Conium Extract, cks........	1		322
Copperas		6372	366
Zinc, White, cks........	6	3919	100
Copper Percipitate	12	14947	118
Cream Tartar, cks........	129	94540	53

Item			
Gum Senegal, bgs........		491	86396
Gum Shellac, cs........	24239	41	10033
" Tragacanth .		6	12464
" bags	33416	11	....
Aden Galls, bags........	1759875	5	413
" " bgs........	44154	50	8487
Indigo, pkgs........	67200	140	....
" pkg	500	351	103137
" Aid Extract cks........		6	5299
" Natural, cks........		12	6875
" Paste, pkgs........		65	44856
Indian Red, kgs........		20	2000
Iodine, cs........		2	300
" Green, cs........		2	200
" cls, cs........		5	500
Isinglass, cs........	145276	34	6725
Jalap, bales........	24850	5	1006
Lac Dye, cs........	1941384	29	6902
Lac Nitrate cf........	5823158	:	16870
Ligumvite........		3	75040
Licorice, cs........	237354	15075	....
Linseed, pkg........		8250	45600
Linseed, pkgs........	1027	156	....
Lac, Chloride, cks........	307115	:	152705
Lina Wood........		50	67200
Logwood, cs........		54	....
" cks........	3247	3	19964000
Mr, cks........		23	57115
" Extract, cs........	2416	15	13
Magnesia, cs........	368005	1	....
" cs........	43543	25	680
" Calci, pkgs........	20756	15	224
" Carb, cs........	22874		3720
" Cit ta, cks........	14584		250
" Citrate, bbs........			

BOSTON IMPORTATIONS CONTINUED.

Item	Lbs.	Pkgs.
Magnesia Sulphate, cks.	5	4700
Manganese, ck.	1	....
Medicinal Preparations, pkgs.	7	...
Mineral Red, pkgs.	30	503
" White, cks.	50	9292
Musk, cs.	1	oz 40
Myraborlan, bgs.	43	8400
Nutgalls, bags.	105	25763
" Mfg, Paste, cz.	3	2500
Ochre, cks.	5	2563
Ochre. ... cks.	2	1190
" in Oil, ck.	1	...
" Oxford, cks.	4	2874
Ochre Paint, cks.	18	22633
Ore Yellow, cks.	49	5287
Oil ... cs.	5	...
" ... cs.	6	1025
" Bergamot, cs.	6	153
" ... cs.	12	...
" Citronela, cs.	108	oz 59520
" Cod Liver, cks.	277	grll 7580
" ... cks	192	...
" ... cks.	5	...
" ... cs.	24	...
" Dog, bbl.	1	...
" Essential, cks.	1	...
" Gallepole, cks.	19	...
" Gioza, cks.	13	...
" Lavender, cs.	2	29
" Lemongrass, ...	7	oz 5038
" Lemon, cans.	7	301
Oil, Lemon, pkgs.	98	...
" ..., cs.	2	...
" Nutmeg, cs.	9	...
" Nut, cs.	1	...
" ..., pkgs.	590	2583
" Orange, cs.	50	...
" Palm, ...	4106	gal 596293
" ... pkgs	1160	...
" Patchouly, cs.	12	...
" Rape, bbls	8	...
" Rapeseed, ck.	1	309
" Sod, cks.	33	3050
" Sperm, cks.	318	63638
" Salad, cks.	220	...
Orange Lead, pkg	86	6470
" ..., cks.	6	...
Oxide Iron, bbls.	400	13440
Oxydising Paste, cks.	7	...
" ... kegs	8	881
Paints, cks.	93	...
" pkgs.	164	35227
Paint Lead, cks.	49	23417
" Black, cks.	5	560
" ... kegs.	10	...
" ..., cks.	82	13481
" Orange, cks.	7	4085
" ... cks.	8	...
" ... cks.	83	39233
Paint Yellow kgs,	110	8378
Paris ... cks.	120	34444
Patent Blue, cs.	2	...
Persian Berries, bugs.	311	78757
Pigment Green, cks.	10	...
" ..., cks.	8	8581
" Brown, cks.	5	4554
" Orange, cks.	7	...
Potash ... cks	224	89701
" cks.	62	42804
" ... of.	75	7840
" Iodine, cs.	8	212
Plaster of Paris, cks.	10	3360
Redwood, pcs.	15261	42564
Red Lead, kgs.	106	...
" pkgs.		1074
Rhubarb, pkg.	1	64
Safflower, Ext. of, cs.	5	125
Sago Pearl, pkgs.	428	700
Salts Aniline, cs.	1	612
Sal Soda, cks.	9068	1335195
Saltpetre, bgs.	1079	...
"		116004
Sapan Wood, cks.	8151	873453
" ... peices.	1	...
Scurlet Poneeau, keg.	37	11743
Senna, ba es.	2	5216
Senna ... bes.	175	33957
Shellac, ckts.	21	23850
Size, cks.	14	...
" cks.	170	76191
Soda, cks.		
Soda Ash, cks.	10364	13145079

	Item			Item			Item		
	Soda, Arsenate of, cks.....	7	13636	Tar, cks..............	11	4336	Varalosi.....	5	1236
	" Bi-carb, pkgs...	10226	4433	" Sal, cks..........	10	5314	Varnish, cans..........	16	gals 384
	" " pkgs...........	993	....	Tapioca, pkgs.........	1191	208586	" cs...............	11	...
	" Carbonate, kgs.........	2522	282446	" bgs.....	765		Vegetable Yellow, cs........	1	...
	" &c, pkgs,.........	2.267	1390591	Tra Alba, cks.........	590	326509	Venetian Red, bbls........	900	568800
	" Hypo cks.............	59	....	Tumeric, bgs.......	941	136406	" cks..........	67	...
	" " cks..........	216	90326	" pkgs..........	182	....	Vermilion, cks..........	2	1030
	" le of, bags.....	21885	918	Ultramarine, pkgs...	155	50901	Water Colors, cks.....	11	....
	" Silicate of, cks...	45	63449	Umber, cks............	59	8813	White Lead not specified, pkg	1194	252743
	" Stanate of, cks...	326	582	" cks........	102		" dry, pkgs........		21237
	Starch Syrup, cks...	75	67857	Van Dyke Ben, cks..........	11	9931	Whiting Gilders.........	...	36087
	Sumac, bgs.............	16725	3133986	" " cks..........	2		Zinc, cks........	258	...
	Superphos Lime, cks..... ...	7	6720	Vanilla Beans, cs..............	1		Zinc Oxide, cks............	20	7544

EXPORTS OF DRUGS, OILS, PAINTS, CHEMICALS, Etc.

—:o:—

The following is a complete list of the exports and re-exports of Drugs, Oils, Paints, Chemicals and Dyestuffs for the year 1873. In instances where the weight cannot be correctly stated the number of packages alone will be given, and also in a few cases where we have been unable to ascertain the number of packages, we give the pounds only.

Article	Pkgs.	Lbs.	Article	Pkgs.	Lbs.	Article	Pkgs.	Lbs.
Acid, Acetic	4	....	Antimony	93	22912	Bark Liquor	9	....
" Carbolic	1	gals 10	Ammonia	51	135	" Peruvian	1061	120954
" Citric	6	568	" Salts	6	292	" Quercitron	2	2
" Muriatic	221	....	Argols	2	....	" "	12487	2753264
" Nitric	35	....	Arrowroot	6	268	" Quilla	1081	....
" Not specified	3246	....	"	5	....	" Quinine	15	5953
" Steric	2	1571	Arsenic	1	815	" Sassafras	10	934
"		400	"	3	....	Barytes	1	50
" Sulphuric	297	....	" powdered	1	232	Berries, Juniper	5	....
" Tartaric	13	....	Ashes	1	77687	Beeswax	1	108
" powdered	8	...	"	128	16900	"	573	142608
Alcohol, bbls.	16137	10118	Balsam, not specified	....	1680	Belladonna Leaves	10	....
Alizarine	47	....	" Canada	21	gal 41	Bleaching Powder	....	17927
Aloes	7	....	" Copaiva	1	2189	Blue Vitriol	11	2138
"	5	3136	" Peru	6	4440	Bone Black	14	9716
Alum	446	8092	Bark	40	159.59	Borax	1	856
" pkgs	354	....	" Barberry	1525	....	"	112	123518
Aniline Colors	8	465	" Eln.	104	103	Brimstone	2703	414316
Aniseed	652	9952	" Larch	1	106	"	74	....
"	54	....	" Liquor	1	90	"	4	1455
Anthracene, tcs...	65	....	"	226	?180	Bromine	6	....
" pkgs...	127	46633				"	878	6035

Qty	Article		
1	Dar Bs		
1	al		
70½	al, tns		
75	Cr		
1	Ids		
10	Carb lie		
25	"		
12	In Black		
215	In Buds		
25	la Seed		
4	la Salt		
1	"		
4	Chr me Green		
26	" Yellow		
191	Ids Ids		
3	Ial		
1	Gs		
29	Cd		
75	"		
83	In Tartar		
5	"		
86	Cummin Seed		
3	"		
2	Cuttlefish		
1	"		
13227	Dyewoods, sticks		
565	" tons		
172	Earth Paint		
40	"		
31	Essences, not specified		
4	" Bay leaves		
4	"		
5	" Maravillosa		
2	"		
1	" Spruce		

Value	Article	Qty
24	Extract Flavine	50
20	" Ec	49
.....	" Hypernia	1
11470	" Logwood	7582
.....	"	12555
.....	" Mr	81
gals 250	"	213
1344	" Quercitron	1675
.....	"	100
2700	Flavine	562
250	"	331
.....	Fla Ms	15
206	Rhs Sulphur	5
1802	Fullers Earth	5
17190	Ric, tns	240
140	"	51
300	Mr tis, age	1607
5530	I elg	714
.....	"	30
73141	Glue	169
.....	"	26
9752	Glycerine	1
.....	"	27
243	Gum Arabic	5
.....	" Danar	2
.....	" Guaiac	75
64750	" Myrrh	80
.....	" Sandrac	10
1240	" Thus	1
300	"	60
.....	Hypernia	20
gross 10	Indigo	200
.....	Insect Powder	41
.....	Isinglass	2
		1

Value	Article	Qty	Value
200	Isinglass	5	.....
2578	Lac Dye	283	230269
53	Lam bulk	704	.....
6412	"	225	200
	Lavender Flowers	2	.....
19120	Licorice Paste	1260	245817
	In, Phosphate	41	11450
59284	" Superphos of.	6	1500
	" Isid	56	10148
37540	"	5	.....
	Lobelia Herb.	3	165
2750	Logwood	543	.....
.....	" tons	501	gal 49
.....	" Liquor	1	20
	Lycopodium	1	.....
	Mace Paste	3	.....
125652	Madder	14	14012
111361	Magnesia	204	8184
	"	2	.....
19802	" Carb	27	2152
	" Citrate	13	.....
50	Maiden Hair.	2	400
	Manganese ..	113	73960
668	Manna	4	.....
416	Mercury	2	180
	"	1	76¼
12293	Ochre	67	8262
2381	"	41	.....
9125	Oil, Almond	122	9092
10463	" Anise	82	.....
5601	" Cajeput	29	1633¾
	" Castor	34	.....
58	"	998	gal 20129
		101	.....

NEW YORK EXPORTS CONTINUED.

Item	No.	Lbs.
Oil, Cd Livr	17	gals 166
"	285	
" Copaiva	4	gals 48
" Cd Seed, bbls	7581	gals 331795
" " "	1664	
" Fh, bls	1083	gals 43462
" " cases	5	gals 45
" " bbls	828	
Lard, "	2768	gals 113072
" a	498	gals 4980
" lbs	1512	
" fin	64	1363
" Lrd	425	gals 9084
"	85	
Lubricating, lbs	12584	gals 4:0706
" cs	748	6584
" "	31	
" lbs	1252	
" Mn	5:3	gals 492
" Ml, cs	35	" 30
" fhe, cs	1	120
" Mt	148	gls 625
Paraffine, Wx		125
Pearl Ash	20	2400
"	263	gals 1808
Palm	168	
"	69	110298
"	2	
" fhe	2:26	gals 117014
"	156	
Permint	200	7508

Item	Pkgs.	Lbs.
Oil, Peppermint	174	.
" Petit Grain	1	69
" Red, bbls	533	
" " bbls	1185	347650
" Rose	1	ozs 200
" Sassafras	33	8495
" "	4	
" Spearmint	1	45
" Sperm	gls 2419	56.
" " Blk	196	
" Spruce	2	gals 91½
" Vacuu:r, bbls	201	gls 62
" Vitriol	84	14634
" "	51	
" Whale	527	gals 796
Roots,	153	
" "	1	
" Wintergreen	3	50
" fla	288	70 88
" fgrn	938	1330052½
"	16	
Paraffine, Wx	1571	463495
Pearl Ash	62	24849
" rfls	11	1122
" Mo	3	
"	337	79157
"	5:	252
Poppy heads	2	
Potash	1032	554013
"	503	
" Bromide	52	2017½

Item	Pkgs.	Lbs.
Potash, Bromide	21	...
" "		516
Potash, Chlorate	12	1344
" Cyanide	5	...
" Iodide	1	25
" Salt	95	10120
" "	12	
" Sulphate	5	600
Mn Ble	15	512
Cfe	100	5411
Mr	213	21008
Cl	25	
Cl	32	oz 3816
Red Lead	13	
"	138	14744
Roots, Gin	1	113
" Cin	2	165
"	1	156
" Ginger	122	21733
" Kp	6	
" Lee	3	
" M	78	59545
" Ml	163	36366
" Plk	2	
" Sin	1551	264300
" Senega	875	
"	529	45:84
" Snake	35	
"	11	774
" Soap	1	
"	423	190000

Article					
Safflower	22	.....			
Saltpetre	718	120989			
"	6	.....			
Salts, Epsom	445	9101			
" Glauber	63	.....			
" Rochelle	10	230			
"	2	250			
"	11	1186½			
Sapanwood	122	.....			
Seeds,	4	379			
"	1				
" Croton	2	250			
" Flax	52	6442			
"	35				
" Mustard	2	370			
Senna	14	3144			
Shellac	13	1785			
"	7	.....			
Sienna	10	1660			
Sila Ash	279	264251			
Soda Ash	12	70289	10	2240	
" H-carb	569	.....	20	1805	
"	12	.....	7	.....	
"	1383	397194	2233	gals 32353	
"	115	.....	95	.....	
"	2	gals 22	1	128	
" Sal	145	73832	2	66	
" White	1	122	2	86	
Spermaceti	1	336	9	438	
"	2239	154318	2	.....	
" Vegetable	1470	.....	107	13309	
"	1	ozs 72	125	66395	
Sugar Lead	2	1327	.....	893	
"	9	1788	948	885	
"	6	.....	62	.....	
Sumac	200	91148	14	4728	
Zinc	70	.....	155	1941	
"	2	.....	.....	480	
Terra Alba	4	599	271	1 94	
Turmeric	1	216	30	9 09	

GEO. F. GANTZ & CO.,

176 DUANE STREET,

ESTABLISHED } 1849

NEW YORK.

{ GEO. F. GANTZ.
{ JOHN M. JONES.
{ ENOS F. JONES.

IMPORTERS,

SHIPPING AND COMMISSION MERCHANTS,

IN

CHEMICALS, OILS, DRUGS,

AND

AMERICAN PRODUCE.

References :

BROWN BROTHERS & CO., Bankers, New York.
DUNCAN, SHERMAN & CO., " "
MORTON, BLISS & CO., "
NATIONAL BANK OF THE REPUBLIC, "
DAVID TAYLOR & SONS, London.

ARTICLES TO WHICH ATTENTION IS GIVEN:

Soda Ash,	Hypo. Sul. Soda,	Pot and Pearl Ashes,
Caustic Soda,	Bleaching Powder,	Indigos,
Sal Soda,	Castile Soap,	Cocoa Nut Oil,
Bi-Carb Soda,	Cream Tartar,	Essential Oils,
Cutch,	Argols,	Terra Alba,
Gambier,	Chlorate Potash,	Senna.

GEORGE F. GANTZ & CO.,

176 DUANE STREET, NEW YORK.

DIXON.

—:o:—

ESTABLISHED, - - - 1827.

—:o:—

PARIS, 1867.

VIENNA, 1873.

Oldest House in the Trade in the Country.

Graphite,

Plumbago,

Black Lead,

IN ORIGINAL PACKAGES FOR THE FINEST USES.

DIXON'S CRUCIBLES.

DIXON'S

Graphite Pencils,

A NEW LEAD PENCIL in five grades of Leads for office use, and ten grades for artists, from the softest up to the hardest, all of one uniform quality, mounted in round and hexagon cedar, beautiful in many styles.

Patent rubber safety attachment to prevent them from falling from the pocket.

These were the only pencils awarded the **GRAND MEDAL FOR PROGRESS** at VIENNA, 1873.

For descriptive lists of our various productions, and for sample of pencils sent by mail without charge, address

THE DIXON CRUCIBLE CO.,

ORESTES CLEVELAND, President.

OFFICES AND WORKS:—233, 235, 237, 239, 241 and 243 Railroad Avenue, and 246, 248, 250, 252, 254 and 256 Wayne St., JERSEY CITY, N.J

MEDAL OF SPECIAL AWARD.

AWARDED TO
C. H. PHILLIPS,
FOR
Milk of Magnesia
1873.

A Mixture of Pure Hydrate of Magnesia and Pure Water.

(BY A NEW PROCESS.)

The risks from using the Calcined Magnesias, on account of their tendency to form concretions in the bowels, giving rise to inflammation and death, have been often demonstrated, and hence various "Fluid Magnesias," consisting of solutions of Magnesia by means of Carbonic Acid, have been put on the market, but all have proved too unstable for general use. From their weakness, too, they have required to be given in such large doses as to give rise to serious trouble, especially in the case of infants.

This preparation, a mixture of **Pure Hydrate of Magnesia** with **Pure Water,** obviates all these difficulties. Its liquid form and freedom from Carbonic Acid establish its claim to be the only safe Magnesia. Its agreeable taste and milk-like smoothness render it valuable in infancy and derangements of the digestive organs of adults. It will be found as an antacid or remedy for heart-burn much superior to the Bi-Carbonates of Potass and Soda, it having been proved that their use is attended with serious disorders to the coats of the stomach. This Magnesia prevents the food of infants souring on the stomach. It is invaluable in complaints of the bladder; regulates the action of the bowels, and is peculiarly adapted for infants as a mild aperient, and for women. In the ordinary cases of Gout and Gravel this solution forms soluble combinations with the uric acid salts, counteracting their injurious tendency when other alkalies and even solid Magnesia have failed. It is a pleasing aperient, and in cases of irritation of the stomach, from excess of eating or drinking, speedy and gentle in removing acidity, relieving and settling that organ in a very short time.

It is esteemed by the profession, and should be kept in all families where there are young children, as they require no persuasion to take it; and for adults it is an agreeable aperient and antacid, removing headaches, from sourness or acidity of the stomach.

THIS MAGNESIA IS PREPARED ONLY BY

C. H. PHILLIPS, New York,

AND SOLD BY ALL DRUGGISTS THROUGHOUT THE UNITED STATES, CANADA AND THE WEST INDIES.

IN BOTTLES, HALF PINTS 50 CENTS; LARGE BOTTLES, $1.

DANIELL & CO., General Agents,

58 Cedar Street, New York.

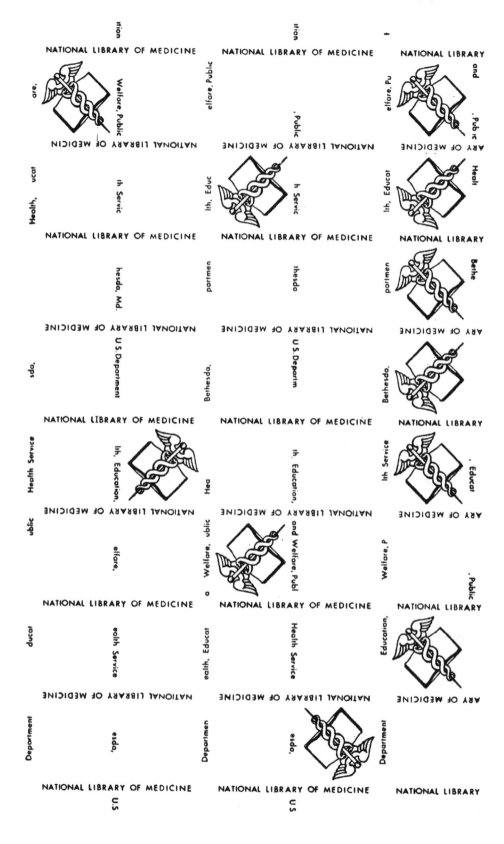

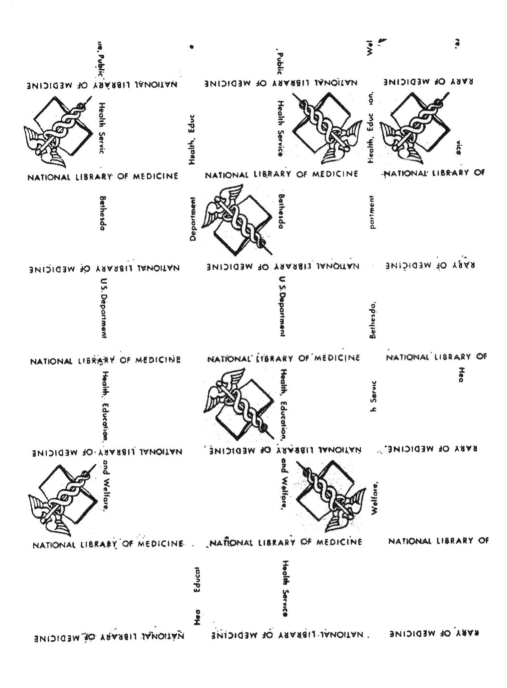

CPSIA information can be obtained
at www.ICGtesting.com
Printed in the USA
BVHW04*1010190918
527934BV00014B/738/P